NATURAL HISTORY
OF THE
INSECTS OF CHINA,
CONTAINING
UPWARDS OF TWO HUNDRED AND TWENTY
FIGURES AND DESCRIPTIONS.

中国昆虫志

［英］爱德华·多诺万　著
［英］约翰·韦斯特伍德　修订

Edward Donovan
John Obadiah Westwood

上海古籍出版社
SHANGHAI CHINESE CLASSICS PUBLISHING HOUSE

Selection of Rare Books from Bibliotheca Zi-ka-wei

徐家汇藏书楼珍稀文献选刊

徐锦华　主编

本书系国家社科基金重大项目
"徐家汇藏书楼珍稀文献整理与研究"
（项目批准号：18ZDA179）成果之一

总 序

董少新

文化交流是双向的，这是文化交流史研究的基本共识。但同时我们也必须认识到，在特定的历史时期，文化交流往往是不平衡的。这种不平衡体现在多个方面，其中就包括文化交流双方输入和输出的文化、知识、思想和物质产品等的数量不平衡，也包括己方文化对对方的影响程度不平衡。研究文化交流的这种不平衡性，考察特定历史时期文化交流双方输出和引进对方文化的数量及影响程度的差异，具有重要的学术意义。这样的研究可以在横向对比中为我们评估双方社会的发展程度、开放与包容性、对外来文化的态度、发展趋势及其原因等问题提供重要的参考。

16世纪以后的中西文化交流是人类历史上最伟大的文化交流之一。它不仅对双方造成了深刻的影响，而且一定程度上也促进了人类的近代化进程。对这一时期的中西文化交流史的研究，中外学界已有的成果可谓汗牛充栋。对前人研究略加梳理我们便不难发现，已有成果中更多的是研究西方文化东渐及其对中国的影响，而对中国文化西传欧洲的历史，尤其是中国文化在欧洲的影响史，虽然也有不少研究，但整体而言仍是远远不够的。这便给我们造成这样一种印象，认为西方先进的科技文化对中国造成了深远影响，而中国落后的农耕文化对西方的输出和影响十分有限；西方带动并主导了近代化进程，中国一度因为闭关锁国而错失了跟上先进的西欧发展步伐，最后不得不在西方的坚船利炮压力下才被迫打开国门，进入世界。导致这样的认知状况的原因很多，也很复杂，其中的一个重要原因是后见之明的影响，即用19世纪中西关系的经验来涵括整个16-20世纪中西关系史。如果我们以公元1800年为大约的分界线，将16世纪以来的中西关系史分为前后两个时期，那么不难看出，很多以往的观点和印象对16-18世纪的中西关系史并不适用，有的甚至是截然相反。

耿昇在法国学者毕诺（Virgile Pinot）《中国对法国哲学思想形成的影响》中译本"译者的话"中说："提起中西哲学思想和科学文化的交流，人们会情不自禁地想到西方对中国的影响。但在17-18世纪，中国对西方的影响可能要比西方对中国的影响大，这一点却很少有人提到。"在17-18世纪，到底是中国对西方的影响大，还是西方对中国的影响大？这是一个很值得思考并需要从多个角度加以回答的问题。

相关文献的数量或许是回答此问题的重要维度。就我个人的研究经验和观察而言，这一时期有关中国的西文文献的数量，要远远超过有关欧洲的中文文献数量。来华的西洋传教士、商人、使节和旅行家根据自己的所见所闻、亲身经历甚至中国典籍，用欧洲文字书写了数量庞大的书信、报告、著作和其他档案资料，绝大部分都被寄送或携带至欧洲，从而将丰富的中国信息传回了欧洲。这一时期曾到过中国的欧洲人数以万计，仅天主教传教士便有千余人，其中不乏长期在华、精通中文者。这些西方人是这一时期中西文化交流的主要媒介，其数量远超曾到过欧洲的中国人，而且黄嘉略、沈福宗、胡若望、黄遏东等少数去过欧洲的中国人，其主要扮演的角色和发挥的作用也是向欧洲传播中国文化和知识。来华传教士，尤其是实行适应性传教策略的耶稣会士，的确用中文翻译、撰写了数百部西学作品，但是数量上远不及他们以西文书写的介绍中国的书信、著作、报告乃至图册。也就是说，这一时期传入欧洲的中国知识和信息远多于传入中国的欧洲知识和信息。如果将带有丰富文化、艺术信息的瓷器、漆器、丝织品、外销画、壁纸、扇子等物质文化商品也考虑进来，中西文化交流在数量上的差距便更为明显，毕竟这一时期欧洲商人带入中国的作为商品的物质文化数量是相当有限的。

另一方面，17-18世纪欧洲的知识界根据来华传教士和商人带回的中国信息、知识而撰写的文章、小册子和书籍，其数量更远远超过中国知识界根据来华传教士和商人带至中国的欧洲信息、知识而撰写的作品。1587年出版的门多萨（Juan Gonsales de Mendoza）《中华大帝国史》（*Historia del Gran Reino de la China*）在欧洲被翻译成多种文字并一版再版，同时期没有一部有关欧洲的中国学者的作品出现；1735年出版的杜赫德（Jean Baptiste du Halde）《中华帝国全志》（*Description géographique, historique, chronologique, politique et physique de l'Empire de la Chine et de la Tartarie chinoise*）同样在欧洲广为流传，但同时期并没有一部中国学者的作品可与其相提并论，即便魏源的《海国图志》或可与之相比，其出版时间也已晚于《中华帝国全志》一个多世纪。从这个角度来看，中国知识和文化在欧洲的影响要远大于欧洲知识和文化在中国的影响。这一点还可以从欧洲盛行一个世纪的"中国风"以及一批启蒙思想家对中国文化的讨论中看出，而这一时期传入中国的欧洲艺术主要局限于清宫之中，从黄宗羲、顾炎武、王夫之到钱大昕、阎若璩、戴震等清代主流学界到底受到西学何种程度的影响，也还是不甚明了的问题。

当然，仅从文献数量来论证中国对欧洲有更大的影响并不充分。关于16-18世纪中国文化在欧洲的传播和影响，一个多世纪以来欧美学界有过不少专门的研究，如艾田蒲（Rene Etiemble）《中国之欧洲》、毕诺《中国对法国哲学思想形成的影响》和拉赫（Donald F. Lach）《欧洲形成中的亚洲》等。这些著作对打破欧洲中心主义的偏见发挥了重要作用，但我们也必须看到，这些著作在欧美学术界是边缘而非主流。在"正统"的欧洲近代史叙述中，欧洲所取得的成就是欧洲人的成就，是欧洲人对人类的贡献，是欧洲自古希腊、罗马时代以来发展的必然结果，包括中国在内的非欧洲世界的贡献及其对欧洲的影响几乎被完全忽视了。

中国学界方面，早在20世纪30-40年代，钱锺书、陈受颐、范存忠、朱谦之等学者便对中国文化在欧洲（尤其是英国）的传播和影响作了开拓性的研究。但此后中国学界对该问题的研究中断了较长的时间，直到20世纪90年代以来才重新受到学界的重视，出现了谈敏《法国重农学派学说的中国渊源》、孟华《伏尔泰与孔子》、许明龙《欧洲十八世纪中国热》、张西平《儒学西传欧洲研究导论：16-18世纪中学西传的轨迹与影响》、吴莉苇《当诺亚方舟遭遇伏羲神农：启蒙时代欧洲的中国上古史论争》、詹向红和张成权合著《中国文化在德国：从莱布尼茨时代到布莱希特时代》等一系列著作。但这些研究主要集中于中国文化对英、法、德三国启蒙思想家的影响，至于中国知识、思想、文化、物质文明、技术、制度等在整个欧洲的传播和影响这个大问题，仍有太多问题和方面未被触及，或者说研究得远非充分。例如，包括中国在内的非欧洲世界的传统知识、技术对欧洲近代科技发展有何种程度的影响？中国、日本、印度、土耳其乃至美洲的物质文化对欧洲社会风尚、习俗、日常生活的变迁起到了什么样的作用？近代欧洲逐渐形成的世俗化、宗教包容性、民主制度除了纵向地从欧洲历史上寻找根源之外，是否也存在横向的全球非欧洲区域的影响？世界近代化进程中，包括中国在内的非欧洲世界以何种方式发挥了怎样的作用？

对于这些问题的研究和讨论，首要的是掌握和分析16世纪以来欧洲向海外扩张过程中所形成的海量以欧洲语言书写的文献资料，其中就包括来华欧洲人所撰写的有关中国的文献，和未曾来华的欧洲人基于传至欧洲的中国信息和知识写成的西文中国文献。在这方面，西方学者比中国学者更有语言和文献学优势，在文献收集方面也拥有更为便利的条件。中国学界若要在中国文化西传欧洲及其影响问题上与欧美学界开展平等对话，乃至能够有所超越，必须首先在语言能力和文献掌握程度上接近或达到欧美学者的同等水准。实现这一目标极为不易，但近些年中国学界出现的一些可喜的变化，使我们对这一目标的实现充满期待，这些变化包括：第一，中西学界的交流越来越频繁和深入；第二，越来越多的年轻学者有留学欧美的经历，掌握一种乃至多种欧洲语言，并对近代欧洲文献有一定程度的了解，具备利用原始文献开展具体问题研究的能力；第三，中国学界、馆藏界和出版界积极推动与中国有关的西文文献的翻译出版，甚至原版影印出版。

上海图书馆徐家汇藏书楼拥有丰富的西文文献馆藏，不仅包括法国耶稣会的旧藏，而且包括近些年购入的瑞典汉学家罗闻达（Björn Löwendahl）藏书。徐家汇藏书楼计划从其馆藏中挑选一批珍贵的西文中国文献影印出版，以方便中国学界的使用。第一批出版的《中国植物志》《中华和印度植物图谱》《中国昆虫志》《中国的建筑、家具、服饰、机械和器皿之设计》《中国建筑》《中国服饰》均为17-19世纪初中西文化交流的重要文本或图册，是研究中国传统动植物知识、建筑、服饰、家具设计等在欧洲的传播和影响的第一手资料。这批西文文献，以及徐家汇藏书楼所藏的其他珍稀西文文献的陆续出版，无疑将推动中国学界在中学西传、中国文化对欧洲的影响等方面的研究。

（作者为复旦大学文史研究院研究员，博士生导师）

导　言

邓　岚

在自然界中，昆虫是数量最多、分布最广的类群。对于人类社会来说，昆虫亦是最为常见的生物之一。数千年来，人类在与昆虫的长期接触中，积累了丰富的经验与认识。在古代中国，从"养蚕取丝""养蜂取蜜"的生产活动到"掘沟灭蝗""烧火驱蝗"的虫害防治，再到"斗蟋取乐"的闲暇娱乐，中华先民们在对昆虫的开发利用与防治中，不断缩小着与昆虫的距离。在古代西方，从亚里士多德（Aristotle）的《动物志》（Περὶ τὰ Ζῷια Ἱστορίων）到老普林尼（Gaius Plinius Secundus）的《博物志》（Naturalis Historia），再到大阿尔伯特（Albertus Magnus）的《论动物》（De Animalibus），西人在收集、观察、描述与分类的博物学传统中，把昆虫写进典籍里，不断更新着对昆虫的认知。

大约从近代开始，西方社会开启了对昆虫科学与系统的研究。这一时期，显微镜的发明与运用放大了昆虫的微小天地；博物学的蓬勃发展，激发了一批又一批博物学家们探索昆虫世界；大航海时代所带来的物种交换，则为散落在世界各地的昆虫创造了被认识的可能。

换而言之，近代以来，西方对昆虫的研究进入了快速发展阶段，"昆虫复眼的发现""昆虫特殊排泄器官的发现"等有关昆虫的新发现不断涌现；法布里丘斯（Johann Christian Fabricius）的《昆虫学系统》（Systema Entomologiae）、威廉·柯尔比（William Kirby）和斯班司（William Spence）的《昆虫学概论》（Introduction to Entomology）等有关昆虫学的研究著作也层出不穷。而伴随着中西交流的深入，西方也出现了与中国相关的昆虫学著作。在这之中，英国人爱德华·多诺万（Edward Donovan）于1798年出版的《中国昆虫志》（An Epitome of the Natural History of the Insects of China）无疑是一部不可忽视的著作。

作为博物学家和博物插画家，多诺万擅长博物绘画且醉心于收集昆虫标本，曾在伦敦创办过自然博物馆与研究所，著有《不列颠昆虫志》（The Natural History of British Insects）、《不列颠珍稀鸟类志》（The Natural History of British Birds）等多部为人熟知的作品。其中，《中国昆虫志》的出版是多诺万在其"图文并茂"的著述基础上，对中国昆虫研究的一次尝试。

在该书中，多诺万按照林奈生物分类法的排列方式，对其所搜集的一百多种中国昆虫信息进行分类阐述。同时，书中配有50幅由多诺万依据相关标本所画的彩色昆虫版图，这些版图着色厚重，绘制细致、精美：有的在一幅版面中刻画了多种昆虫；有的用植物为衬托，单独以某种昆虫成画；有的绘画了同一种昆虫在不同视角下的形态。由于书中所收的中国昆虫种类丰富、新奇少见，且插图兼具科学性与艺术性，该书一经出版就引起了公众的关注，取得了不错的反响。随后，多诺万又出版了《印度昆虫志》（*An Epitome of the Natural History of the Insects of India*）及《新荷兰等地昆虫志》（*An Epitome of the Natural History of the Insects of New Holland, New Zealand, New Guinea, Otaheite, and Other Islands in the Indian, Southern, and Pacific Oceans*），它们与《中国昆虫志》一同被认为是多诺万有关域外昆虫研究最为成功的著作。

然而，与那个时代的许多博物学家一样，多诺万从未到过中国考察。他在书中所列举的昆虫信息，或来源于他人在中国自然考察时所获得的资料，如马戛尔尼使团在访华途中所采集的昆虫标本和相关报告；或来源于私人的来自中国的藏品与资料，如珠宝商约翰·弗朗西永（John Francillon）和昆虫学家德鲁·德鲁里（Dru Drury）所收藏的昆虫标本以及博物学家约瑟夫·班克斯（Joseph Banks）的藏品与资料等。

1838年，英国昆虫学家、林奈学会会员约翰·韦斯特伍德（John Obadiah Westwood，1805-1893）根据当时昆虫学研究所取得的最新进展，对该书进行了修订再版（*Natural History of the Insects of China*）。本书此次影印的底本即为目前徐家汇藏书楼所藏该书的1842年修订版。

与1798年初版相比，在该书的1842年修订版中可以看到，韦斯特伍德对1798年版中所出现的错误说法（如地名等）做了修改，并在部分正文或脚注处增添了新的看法与解释。此外，韦斯特伍德还为书中的版画及版画中所绘的昆虫进行编号，使其阅读与查找更为清晰、方便。同时，此本还对文末的索引表做了改变。在1798年版中，该书附有一份以法布里丘斯的昆虫分类标准排列的索引表及一份以林奈的昆虫分类法排列的索引表。而在此版中，文末替换为一份按昆虫名称的字母顺序排列的索引表和一份依据彼时昆虫系统分类所编排的索引表。

值得一提的是，据韦斯特伍德在此版的序文中所言，该书的德文版曾于1801年由约翰·戈特弗里德·格鲁伯（J. G. Gruber）在莱比锡编辑出版，但由于未亲眼见过此本，韦斯特伍德也无法确定该版为节译本还是全译本。因此，在目前已知的该书所有版本中，由韦斯特伍德修订出版的《中国昆虫志》最为全面、准确与专业。

作为一本有关中国昆虫研究的西方著作，《中国昆虫志》的撰著和出版有着独特的价值与意义。一方面，该书撰著于西方昆虫学研究的上升阶段，是当时西方昆虫学发展下的必然结果，也是当时西方昆虫学研究所取得的成就体现。从1798年初版到1842年修订版，透过精美的昆虫绘图与愈发完善的专业阐释，该书版本变化的背后隐藏着的是西方自然科学发展的轨迹。

另一方面，《中国昆虫志》是中西交流下的产物，为当时西方认识中国提供了自然科学视野。在鸦片战争之前，除了少数传教士、商人、使团之外，欧洲人很少有机会来到中国，更无法进行大范围的自然考察，其对中国自然的认知多来源于上述群体从中国发回的报告、信件、书籍以及商贸货品。即便此时已有部分西人意识到中国拥有丰富的自然资源且极具研究意义，但是西方社会对中国自然的认识及研究依然有限。正如多诺万在《中国昆虫志》初版的序言中所写："如果说中国的自然物产受到人们的关注较少，那只是因为它们的价值和重要性鲜为人知；如果认识得更深，它们会立即使我们惊讶，并使我们相信它们的效用。"

因而与19世纪中叶后西方日渐增多的有关中国昆虫研究的书籍相比，出版于18世纪末的《中国昆虫志》显然具备一定的领先性与珍稀性。或许当时西人从中国寄回的昆虫标本、生物绘图乃至此书的撰著带有一定的"海外探索热情"与"帝国野心"，但就西方认识中国的昆虫种类及运用来看，该书架起了中西之间的认识桥梁，是重要的文献资料。而对我们而言，此书以"他者"的视角，为我们保留了其时其地的昆虫研究资料，对我们了解那一时期中国的物种情况、认识中西交流下的历史发展，具有重要的史料价值。

总而言之，人类文明的发展与交流常伴随着认识的改变，这是一种从模糊到清晰的双向或多向的过程。在这个过程中，书籍既是关键的知识传播媒介，也可能是最好的亲历者与见证者。当我们以现今的科学眼光重新审视此书时，书中的相关信息也许存在着诸多错误，但若想要回顾那一时期的西方自然科学发展、中西交流，该书则于中国、于西方甚至于世界而言，都值得再次被翻阅。

（作者为上海图书馆历史文献中心馆员）

NATURAL HISTORY OF THE INSECTS OF CHINA

CHINA BRANCH
of the
ROYAL ASIATIC SOCIETY.
Shanghai.
Library No. ....

NOT TO BE
TAKEN AWAY

1022

中国昆虫志

NATURAL HISTORY

OF THE

INSECTS OF CHINA,

CONTAINING

UPWARDS OF TWO HUNDRED AND TWENTY

FIGURES AND DESCRIPTIONS,

BY

E. DONOVAN, F.L.S. & W.S.

A NEW EDITION,

BROUGHT DOWN TO THE PRESENT STATE OF THE SCIENCE,
WITH SYSTEMATIC CHARACTERS OF EACH SPECIES, SYNONYMS, INDEXES, AND OTHER ADDITIONAL MATTER,

BY J. O. WESTWOOD,

SECRETARY OF THE ENTOMOLOGICAL SOCIETY OF LONDON, HON. MEM. OF THE LITERARY AND HISTORICAL SOCIETY OF QUEBEC, AND OF THE NATURAL HISTORY SOCIETIES OF MOSCOW, LILLE, MAURITIUS, ETC.

LONDON:
HENRY G. BOHN, YORK STREET, COVENT GARDEN.
MDCCCXLII.

PREFACE.

In presenting a new edition of the Epitome of the Insects of China to the entomological public, I have endeavoured to bring it down to the present state of the science. The former edition, like all the writings of Donovan, was arranged in accordance with the system of Linnæus, and bore the date of 1798. At that period the science of Entomology was in its infancy; but in the subsequent forty years the progress which has been made, has indeed been rapid. I have endeavoured to render the specific characters more precise, the nomenclature more correct (giving the priority to the oldest specific name), and the synonyms more numerous. The localities in many instances were incorrectly given in the former edition; and I have added many additional observations, either incorporated in the text or given as foot notes, omitting nothing which appeared in the former at all likely to instruct or interest the reader. Alphabetical and systematic indices have also been introduced. I dare not hope that this edition is faultless: I have endeavoured to render the beautiful figures of Donovan as serviceable as possible, and must trust to the indulgence of the more skilful specific entomologist. One circumstance may be mentioned which will, at all events, be deemed an improvement, namely, the introduction of numbers both for the plates and for the several figures on each plate. Those who have consulted synonymical authorities in Entomology are aware of the trouble and confusion which have originated in the want of a

A

PREFACE.

regular numeration of these plates, which was caused by the original periodical appearance of the work, whence it happens that we are referred to Part II. plate 1. (as it may be), and not to a consecutive series of numbers upon the plates, which were, indeed, entirely without a number, and appeared promiscuously.

A German edition of this work was commenced at Leipzic in 1801, edited by J. G. Grubner, but I am not certain whether the entire work was republished or only one of the parts.

Of the Entomology of China little more is known at the present time than Donovan was acquainted with. It is true we continue to receive numerous boxes of insects from China, chiefly purchased in the shops of Canton, but, like every thing Chinese, there is such an absolute monotony in these arrivals, that it is almost impossible to discover in a quantity of these boxes a single species which is not contained in all the rest. It is evident that a considerable employment is produced by the rearing of the Atlas moth and some other species, and in the collection of the other insects which we receive in such abundance. The Chinese boxes are made of a soft wood, about 16 inches by 11 in size, and of a sufficient depth to admit a tall needle; a layer of butterflies and moths stuck close to the point of the needle is placed at the bottom of the box, with another layer of beetles, flies, &c. closely packed together and stuck high up on the needles, the points of which are passed through the wings of the butterflies forming the under layer.

Donovan well observed that "the Chinese, like their neighbours the Japanese, are well acquainted with the natural productions of their empire, and Zoology and Botany, in particular, are favourite studies amongst them."

That the Chinese also pay considerable attention to Entomology is evident, not only from the fact of the employment of silk having had its origin in that country, but also from the numerous beautiful drawings of insects upon rice

PREFACE.

paper, brought to Europe in great quantities. Many of these figures are, however, evidently fictitious, although some are occasionally found accurately correct and most elaborately pencilled.

The few but interesting hints which Sir George Staunton (who accompanied the embassy of the Earl Macartney to China) has given on the practical Entomology of China in the account of his Travels, which was published shortly before the appearance of the first edition of this work, were embodied by Donovan in his pages, and from whence we may be induced to hope that at some future time, some of the insects, as well as plants, of that vast empire may become no less objects of utility and importance than of curiosity; the Chinese cochineal insect,* and that from which the wax of the East is procured, are two species that deserve particular attention. The medical precepts of the Chinese will certainly find but few votaries in Europe, but as articles of medicine, amongst others, the Mylabris Cichorii, regarded as the Cantharides of the ancients and still used as a vesicant by the Chinese, may be of importance, as it is said to possess more virtues than the Cantharis vesicatoria of Europe.

From the vast extent of the Chinese empire and our comparative ignorance of its insect productions, it is almost impossible to speak with any precision upon the interesting subject of its entomological geography. Many of its insects bear a great resemblance, and occasionally appear identical with those

---

* Dr. Anderson found eight species of Coccus at Madras. One of these, he says, was found on a young citron-tree, citrus sinensis, just landed from China; it was more deeply intersected between the abdominal rings than any of those of the coast, and he therefore named it C. Diacopeis. *Collection of Letters from Madras, January* 28, 1788. The Cactus Cochinillifer had been found previous to the appearance of the first edition by Mr. Kincaid at Canton; its Chinese name is Pau wang. This had been transmitted to the Nopalry of the Hon. East India Company at Madras, and promised to be of future advantage to the commercial concerns of Great Britain.

PREFACE.

of the eastern parts of India. These are, therefore, doubtless the inhabitants of the immediate neighbourhood of Canton, which lies in the same degree of latitude with Bengal. A valuable addition has indeed been made to our knowledge of the entomological productions of Mongolia, and some of the adjacent northern portions of China, by M. Faldermann's publication* of the Coleopterous insects brought from those regions by M. Bungius; from which it would appear that the entomological productions of such portions of China as lie between 40° and 50° of north latitude are quite unlike those of the neighbourhood of Canton lying beneath the tropic of Cancer, bearing, indeed, a far greater resemblance to those of central Russia. It is to be hoped that M. Faldermann will speedily publish the descriptions of the species belonging to the other orders of insects brought from this interesting portion of the globe.

J. O. W.

---

* Coleopterorum ab illustriss. Bongio in China boreali, Mongolia et montibus Altaicis collect. et ab illustr. Turczamnoffio et Sichuckino e provincia Irkutzh missorum. (In Mem. Acad. Imper. des Sciences de St. Petersbourg, Tom. 2. 1835.)

1. Copris Midas.   2. Oryctes Rhinoceros?
3. Onticellus cinctus.   4. Scarabæus sanctus?
5. Gymnopleurus sinuatus.

INSECTS OF CHINA.

Order. COLEOPTERA. *Linnæus.*

COPRIS MIDAS.

Plate 1. fig. 1.

TRIBE.	LAMELLICORNES, *Latreille.* (Genus, Scarabæus, *Linnæus.*)
FAMILY.	SCARABÆIDÆ, *Mac Leay.*
GENUS.	COPRIS, *Geoffroy.* (Scarabæus p. *Donovan.*)
CH. SP.	C. thorace retuso tricorni, cornu intermedio lato emarginato, capitis clypeo 3-sinuato denteque utrinque laterali valido recurvo armato. Long. Corp. 1½ unc. C. with the thorax blunt in front and armed with three horns, the middle one broad and notched at the tip, the shield of the head with three sinuations, and armed on each side with a thick horn. Length 1½ inch.
SYN.	Scarabæus Midas, *Fabricius Ent. Syst.* I. p. 45. *Syst. Eleuth.* I. p. 36. *Oliv. Ent.* I. t. 30. f. 183.

THE original description of Fabricius was taken from a specimen in the Banksian cabinet, the locality of America being assigned to it; some mistake must, however, have occurred in this respect, as it is now well known that East India is the true country of this curious species, which has received its specific name, Midas, from the size of the ear-like pair of horns at the sides of the head. The accompanying figure was taken from a specimen in the collection of Drury, which was said to have been received from China. Some interesting observations, on the habits of this species, have been published by Col. Sykes in the first volume of the Transactions of the Entomological Society of London.

COLEOPTERA.

ORYCTES RHINOCEROS?

Plate 1. fig. 2.

FAMILY. DYNASTIDÆ, *Mac Leay.*
GENUS. ORYCTES, *Illiger.* Scarabæus p. *Linn.*
CH. SP. O. thorace subbituberculato, capitis cornu simplici, clypeo bifido, elytris punctatis. Long. Corp. 1⅖ unc.
O. with the thorax depressed in front, and with the rudiments of two tubercles, head with a single horn, clypeus bifid, elytra punctate. Length 1 inch, 5 lines.
SYN. Scarabæus Rhinoceros, *Linn. Syst. Nat.* I. II. *p.* 544. *Fabricius Syst. Eleuth.* 1. *p.* 14. *Röesel. Ins.* II. *Scarab.* 1. *tab.* A. *f.* 7.
Scarabæus Nasicornis, *Donovan*, 1st edit.

Donovan observed upon this species, which he considered as identical with the European Oryctes nasicornis, that " the male is furnished with a long recurved horn on the head: the female has only a small rising on that part. It is found in Europe as well as China." It is evident, however, from the form of the thorax, and striation of the elytra of Donovan's figure, that he had confounded two distinct species. I have little doubt that the insect here figured is intended for the Oryctes Rhinoceros, although the elytra are not represented as being punctured.

ONITICELLUS CINCTUS.

Plate 1. fig. 3.

FAMILY. SCARABÆIDÆ, *Mac Leay.*
GENUS. ONITICELLUS, *Zeigler.* Scarabæus p. *Fabricius.*
CH. SP. On. niger, elytrorum margine pallido, clypeo emarginato. Long. Corp. lin. 4½.
On. black with the margin of the elytra pale coloured, the clypeus notched. Length 4½ lines.
SYN. Scarabæus cinctus, *Fabricius Ent. Syst.* 1. *p.* 69. *Herbst. Col.* II. *p.* 327. *n.* 215. *Oliv. Ent.* 1. 3. 169. 209. *t.* 10. *f.* 90.
Onitis cinctus, *Schonh. Syn. Ins.* 1. *p.* 33. *no.* 20.

SCARABÆUS (HELIOCANTHARUS) SANCTUS?

Plate 1. fig. 4.

Genus.	Scarabæus, *Linn. Mac Leay.* Ateuchus, *Latreille, Dejean.*
(Sub-Gen.	Heliocantharus, *Mac Leay.*)
Ch. Sp.	Sc. cupreo-nitens, clypeo 6-dentato, thorace serrato, elytris striatis. Long. Corp. 1 unc.
	Sc. black slightly shining with brass or copper, clypeus with six teeth, thorax with the margins serrated, elytra striated. Length 1 inch.
Syn.	Copris sanctus? *Fabr. Suppl. Ent. Syst. p.* 34.
	Scarabæus sacer, *Donov.* 1st edit.
	Scarabæus sanctus, *Mac Leay, Horæ Ent. p.* 500, var. ♂?

The insect here figured was regarded by Donovan as identical with the Scarabæus sacer of Linnæus, a species inhabiting the south of Europe. "It is," he observes, "a native of China, and is also found in other parts of the East Indies, in Egypt, Barbary, the Cape of Good Hope, and other countries of Africa, and throughout the south of Europe." The insects inhabiting these various countries are now ascertained to be specifically distinct; so that the reference of the species here figured to the Scarabæus sacer cannot be adopted, and it is not improbable that it is identical with the sanctus of Fabricius.

This tribe of insects is especially interesting from its containing the sacred beetle of the Egyptians, by whom it was regarded as a visible deity; but a more refined system of religious worship prevailed in their temples among the priests and sages. They deemed it only the symbol of their god, and, ascribing both sexes to the beetle, it became a striking emblem of a self-created and supreme first cause.*

This insect was more especially the symbol of their god Neith, whose attribute was power supreme in governing the works of creation, and whose glory was increased, rather than diminished, by the presence of a superior being, *Phtha*, the creator. The theological definition of the two powers, being independent, yet centering in one spirit, is implied by the figurative union of two sexes in the beetle. In the latter sense it signified therefore but one omnipotent power. The Scarabæus, typifying Neith, was carved or painted on

---

* "The father, mother, male and female art thou." *Synesius. Hymn. Phtha.*—"The Egyptian spirit *Phtha* gave chaos form, and then created all things." *Jamblichus de Mysteriis,* sect. 8.

a ring, and worn by the soldiers, as a token of homage to that power who disposed of the fate of battles;* and sculptured on astronomical tables, or on columns,† it expressed the divine wisdom which regulates the universe and enlightens man.

---

* Authors quote a doubtful passage in *Horapollo Hieroglyph. lib.* 1. to support this opinion. That such rings were worn by the ancient Egyptians is beyond conjecture, many remains of them, and some very perfect, have been found in the subterranean caverns and sepulchres in the Plain of Mummies near Saccara and Giza. Those which we have examined are remarkable for the convexity or full *relievo* of the figure sculptured on them; in some it is of the natural size of the insect, but generally smaller; the stone, cornelian, without a rim, and turning on a swivel ring of gold.

† Linnæus says the Scarabæus sacer is sculptured on the antique Egyptian columns in Rome. " Hic in columnis antiquis Romæ exsculptus ab Ægyptiis." *Syst. Nat.* Does Linnæus allude to any remains of those colossal obelisks which Augustus transported to Rome when he subjugated Egypt, or others of more recent date? It would increase the interest of our inquiries to learn that the Scarabæus was among the hieroglyphics, on the two very ancient obelisks, carried from Heliopolis, the city of the Sun.

We are informed by ancient writers, that the Scarabæus engraved on the astronomical tables of these people, implied the divine Wisdom which governed the motion and order of the celestial bodies; that those tables were huge and massy stones or columns of granite, with the characters and figures large and highly embossed; in short, such as were supposed capable of long resistance to the corroding hand of time. Among those the Scarabæus was probably the most conspicuous, its size gigantic, and the figure frequently repeated; for this we have observed even on small Egyptian antiques.

Various valuable remains of tablets, with figures of the *Scarabæus sacer*, are preserved in the British Museum and other collections of antiquities in this country. Those we have examined are of various descriptions, some smaller than the insect itself, others of a monstrous size. The stones on which they are sculptured generally *green nephritic* or *jade stone*, or a kind of *basaltes*, and black marble; the figure *basso relievo* on a tablet or slab, but oftener in *relievo*, with the prominent characters of the insect very accurately defined, particularly the six dentations of the clypeus and those of the tibiæ. The reverse of the embossed side is flat and smooth, and abounds in characters altogether unknown, though, from the number of religious objects of worship occasionally interspersed, we may presume they contain an ample store of the ancient sacerdotal language: the most remarkable were the scarabæus, the sceptre and eye (Osiris), the human figure with a dog's head (Anubis), the hawk (Horus), and the Ibis, or sacred bird. On the thorax of one fine specimen we remarked four elegant figures. One of them is holding a *cornucopia* in the left hand, and a branch in the right: this is perhaps a subordinate deity of the Nile, that river having been once found depictured on an antique Alexandrian coin, like an aged man, holding the cornucopia, and a branch of the Papyrus; denoting its abundance and produce. (In many of the mummies which have been recently unrolled in this country, carved Scarabæi have been found in various positions, especially upon the eye and breast of the body.—*See Pettigrew's History of Mummies.* J. O. W.)

The digression on the mythological history of this insect may be considered by some as a tedious deviation from the pursuit of the naturalist; with others we trust it will be more favourably received; for it proves to the unprejudiced mind how deeply the history of nature, and in the present instance the science of Entomology, involves a most important enquiry into the first philosophical opinions of the human race. The means, however trifling, must not be contemned, which illumine the most sublime of all human researches—*The Study of Mankind.*

1. Copris Molossus.   2. Onthophagus Seniculus.
3. Copris Bucephalus.

COLEOPTERA.

GYMNOPLEURUS SINUATUS.

Plate 1. fig. 5.

GENUS. GYMNOPLEURUS, *Illiger.* Ateuchus p. *Fabricius.*

CH. SP. Gymn. clypeo emarginato, niger subcupreus, antennarum apice flavo. Long. Corp. lin. 8.

G. with the clypeus notched, black with a slight coppery tinge, tips of the antennæ yellow. Length two-thirds of an inch.

SYN. Ateuchus sinuatus, *Fabricius Syst. Eleuth.* 1. *p.* 60. *Oliv. Ins.* 1. 3. *tab.* 10. *f.* 90. *tab.* 21. *fig.* 189.
Scarabæus Leei, *Donovan,* 1*st edit.*

The Scarabæus Leei of Fabricius (to which Donovan referred the insect here figured) is totally distinct, and is identical with the Scarabæus fulgidus of Olivier. The original specimen, described by Fabricius, from the collection of the late Mr. Lee, is now in the collection of the Rev. F. W. Hope.

COPRIS MOLOSSUS.

Plate 2. fig. 1.

GENUS. COPRIS, *Geoffroy.* Scarabæus, *Linnæus.*

CH. SP. C. niger, thorace punctatissimo, retuso, bidentato, utrinque impresso; clypeo lunato, ♂ unicorni integro, elytris lævibus. Long. Corp. 1 unc. 4 lin.

C. black, thorax very much punctured, retuse in front and bidentate, with two lateral impressions, clypeus lunate, the male having a single erect horn, elytra smooth. Length $1\frac{1}{3}$ inch.

SYN. Scarabæus Molossus, *Linn. Syst. Nat.* 1. 11. *p.* 543. *No.* 8. *Fabricius Syst. Eleuth.* 1. *p.* 42. *Herbst. Col.* 11. *p.* 178. *t.* 14. *f.* 1. *Oliv. Ent.* 3. *p.* 100. *var. c. t.* 5. *f.* 37. ♂. *Drury Ins. Pl.* 32. *f.* 2. *2nd edit. p.* 64.

S. Molossus and S. Bucephalus are very common in China. The first seems a local species; the latter is said to be found in other parts of the East Indies. Olivier has given three varieties of Scarabæus Molossus. The specimen figured in the annexed plate is the *var. c.* of that author.

The larvæ of the larger kinds of coleopterous insects, abounding in unctuous moisture, are not less esteemed as food among some modern nations, than they were by the epicures of antiquity. In Jamaica and other islands in the West Indies, the larva of the Prionus damicornis, or Macokko beetle, is an article of luxurious food; and in China many insects in that state are appropriated to the same purpose. Thus, also, the Romans introduced

COLEOPTERA.

the larvæ of the Lucani and Cerambyes in their voluptuous repasts; previously feeding them on farinaceous substances to give consistence to the animal juices.

The learned author of the last account we have of China, says, "Under the roots of the canes is found a large white grub, which being fried in oil is eaten as a dainty by the Chinese." Donovan suggests that perhaps this is the larva of Scarabæus Molossus, which, like many other of the Scarabæi,* may live sedentary in the ground, and subsist on the roots of plants: the general description and abundance of this insect in China favours such opinion. The same author observes in another part of his work, that "the aurelias of the silk worm which is cultivated in China, after the silk is wound off, furnish an article for the table." This also is a very ancient custom among the Asiatics, and even Europeans before the sixteenth century, if we may credit Aldrovandus:† it is certain the worms, if not the chrysalides, were administered in medicine in early ages.‡ Fabricius also expressly states that the insect here figured is medicinally employed in China.

ONTHOPHAGUS SENICULUS.

Plate 2. fig. 2ª. and 2ᵇ.

GENUS.	ONTHOPHAGUS, *Latreille.* Scarabæus p. *Fabricius, Donovan.*
CH. SP.	O. thorace antice, clypeo postice bicorni, elytris substriatis strigis duabus baseos e punctis ferrugineis, punctoque uno alterove flavescente apicis. Long. Corp. lin. 5½.
	O. with the thorax in front, and the hinder part of the clypeus with two horns, elytra slightly striated, with two rows of basal pale spots and with two yellowish apical spots. Length nearly half an inch.
SYN.	Scarabæus seniculus, *Fab. Ent. Syst.* 1. *p.* 43. 142. *Oliv. Ent.* 1. 3. *p.* 124. *t.* 7. *f.* 56. *a. b Panzer in Naturforscher,* 24. *t.* 1. *f.* 5.
	♀ Scarabæus brevipes, *Herbst. Archiv. t.* 19. *f.* 16.
HABITAT.	Tranquebar (*Fabricius*), China, *Donovan, Weber MSS.*

The annexed figures exhibit the two sexes of Scarabæus Seniculus. In some specimens the spots are very indistinct and reddish; in others the wing-cases have faint red striæ. The female has the rudiments of horns on the thorax.

---

* The larvæ of the Scarabæi live in the trunks of decayed trees, in putrid and filthy animal substances, or in the earth. The last are the most injurious, because they destroy the roots of plants. All the known kinds of these larvæ are of an unwieldy form and whitish colour, the skin free from hairs, and only the head and fore feet defended with a shelly covering. (As it is most probable that the habits of this large Copris are analogous to those of the English C. lunaris, I should be rather inclined to regard the cane grub mentioned in the above extract as the larva of a Calandra. J. O. W.)

† The German soldiers sometimes fry and eat silk worms. *Aldrov.*

‡ Silk worms dried, powdered, and put on the crown of the head, help the *vertigo* and *convulsions*; mundify or cleanse the blood, &c. &c. *Schroderus, Serapio, &c. &c.*

1. *Cetonia Chinensis.* 2. *Euchlora viridis.*

COLEOPTERA.

COPRIS BUCEPHALUS.

Plate 2. fig. 3.

GENUS.   COPRIS, *Geoffroy.* Scarabæus p. *Linn. &c.*
CH. SP.   C. thorace retuso quadridentato, capitis clypeo angulato, cornu ♂ brevi erecto haud emarginato, ♀ breviori truncato subemarginato. Long. Corp. unc. 1, lin. 8.
          C. with the thorax retuse in front and 4-toothed, clypeus angulated, the male with a short erect horn not notched at the tip, the female with a much shorter and truncated horn. Length $1\frac{2}{3}$ inch.
SYN.   Scarabæus Bucephalus, *Fabricius Ent. Syst.* 1. *p.* 51. 166. *Herbst. Col.* II. *p.* 174. *t.* 13. *f.* 1. 2. *Oliv.* I. 3. 99. *t.* 4. *f.* 26. *t.* 10. *f.* 92. *t.* 22. *f.* 92. ♀

This species has been confounded with C. Molossus. Both species are here represented on the same plate, in order that the difference between them may be precisely observed.

CETONIA (TETRAGONA) CHINENSIS.

Plate 3. fig. 1 and 1ᴬ.

FAMILY.   CETONIIDÆ.
GENUS.   CETONIA, *Fabricius.* Scarabæus p. *Linnæus, Donovan.*
SUB-GEN.   TETRAGONA, *Gory.*
CH. SP.   C. viridi-ænea, clypeo emarginato, thorace posticè lobato, elytris acuminatis, corpore subtus pedibusque subspinosis, castaneis, tarsis nigris. Long. Corp. unc. 1, lin. 6.
          C. brassy green, clypeus notched and slightly spined, thorax lobed behind, elytra acuminated, body beneath, with the legs, chestnut-coloured, tarsi black. Length $1\frac{1}{2}$ inch.
SYN.   Scarabæus Chinensis, *Forster Cent. Ins.* I. *p.* 2. 2.
          Cetonia Chinensis, *Fabr. Ent. Syst.* I. II. *p.* 126. *b. Oliv. Ent.* I. 6. *p.* 11. *t.* 2. *f.* 5. *a. b. Herbst. Col.* III. *p.* 199. 2. *t.* 28. *f.* 2.
          Scarabæus oblongus, *Brown Illustr. t.* 49. *f.* 4.
          Smaragdinus major, *Voet. Col. t.* 5. *f.* 40.

COLEOPTERA.

EUCHLORA VIRIDIS.

Plate 3. fig. 2.

FAMILY.   MELOLONTHIDÆ, *Mac Leay.*
GENUS.   EUCHLORA, *Mac Leay.*   Scarabæus p. *Linnæus, &c.*
CH. SP.   E. supra viridis subtilissime punctata, subtus cum pedibus aurea vel cupreo-ænea. Long. Corp. lin. 11.
         E. green above, and very finely punctured, body beneath with the legs golden or coppery bronze coloured. Length nearly 1 inch.
SYN.   Melolontha viridis, *Fabricius Ent. Syst.* I. II. *p.* 160. *no.* 23.   *Oliv. Ent.* 1. 5. *p.* 29. 31. *t.* 3. *f.* 21.   *Herbst. Col.* III. *p.* 149, *t.* 26. *f.* 5.   *Mac Leay, Horæ Entomol.* 1. *p.* 148. (Euchlora v.)

RHINASTUS STERNICORNIS.

Plate 4. fig. 1.

TRIBE.   RHYNCOPHORA, *Latreille.*
FAMILY.   CURCULIONIDÆ, *Leach.*
DIVISION.   CHOLIDES, *Schonherr.*
GENUS.   RHINASTUS, *Schonherr.*   Cholus p. *Germar.*   Curculio p. *Donovan.*
CH. SP.   Rh. longirostris, femoribus dentatis, corpore polline flavescente obtecto lateribus nigris, rostro utrinque spinoso. Long. Corp. (sine rostro) lin. 11.
         Rh. with the rostrum long, with a short spine on each side at the tip, thighs toothed, body clothed with a fine yellow powder, except a black stripe on the sides of the thorax and elytra. Length (without the rostrum) nearly 1 inch.
SYN.   Rhinastus sternicornis, *Schonherr. Syst. Curcul. vol.* 3. *p.* 558.
         Cholus sternicornis, *Germar. Ins. spec.* 1. *p.* 214. *t.* 1. *f.* 4.
         Curculio Chinensis, *Donovan,* 1st edit.

This insect seems nearly allied to *Curculio mucoreus*, an Indian species, described by Linnæus, but not figured by any author: the lateral stripe of black, and the denticulations on the posterior thighs of our insect clearly remove it, however, from the Linnæan species.

Except the lateral black stripes, and the rostrum, this insect is totally covered with a bright brown powder, or rather, with very minute hairs which adhere but slightly, and resemble that substance. We observe a similar farinaceous appearance on the *Curculiones, Lacteus, Niveus, &c.* and especially on that gigantic beetle *Scarabæus Elephas.*

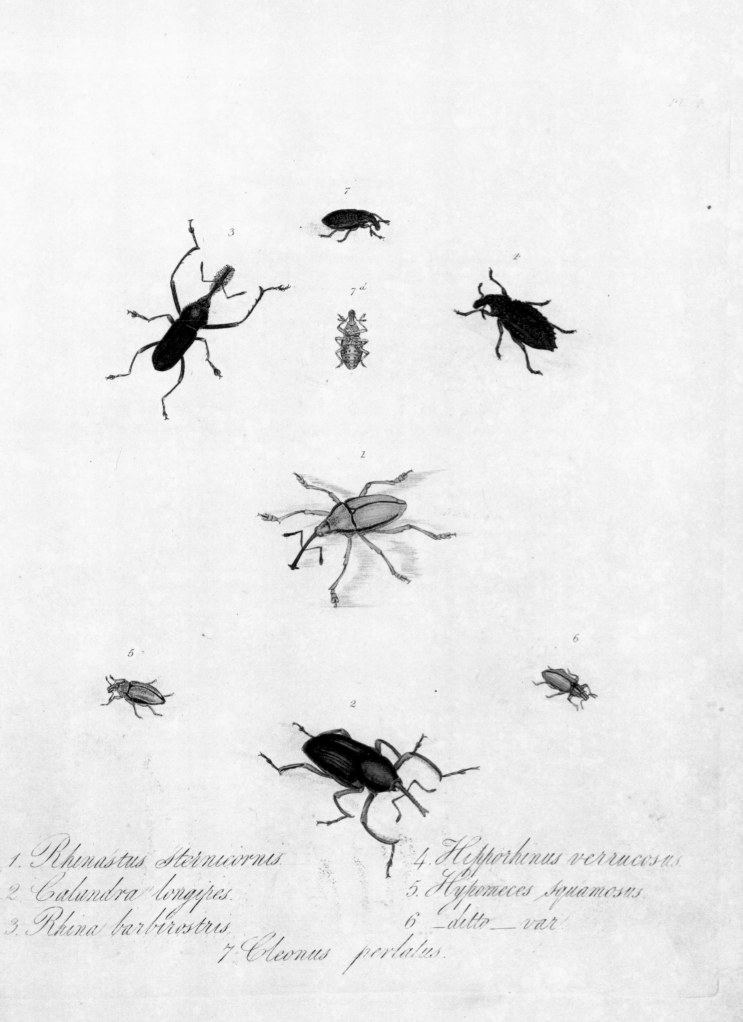

1. Rhinastus Sternicornis.
2. Calandra longipes.
3. Rhina barbirostris.
4. Hipporhinus verrucosus.
5. Hypomeces squamosus.
6. _ditto_ var.
7. Cleonus perlatus.

COLEOPTERA.

I have not the least doubt that this insect is identical with the Brazilian species, Rhinastus sternicornis, of Schonherr, although the specific name given to it by Donovan implies it to be a native of China. Some mistake had doubtless arisen relative to the original country of which it is an inhabitant which led our author into this error

CALANDRA LONGIPES.

Plate 4. fig. 2.

DIVISION.	RHYNCOPHORIDES, *Schonherr*.
GENUS.	CALANDRA, *Fabricius*. Rhyncophorus, *Herbst*. Curculio p. *Linn, &c.*
CH. SP.	C. nigricans; elytris ferrugineis, rostro apice utrinque reflexo (in uno sexu dorso bi serrato,) pedibus anticis elongatis. Long. Corp. (excl. rostro lin. 13.)
	C. blackish-brown, elytra ferruginous; rostrum, with the apex on each side reflexed, and in one sex with the upper edge doubly serrated, the fore legs long. Length 13 lines.
SYN.	Curculio longipes, *Fabricius, Ent. Syst.* 2. 395. *Syst. Eleuth.* 2. 431. *Oliv. Ins.* 83. *tab.* 86. *fig.* 191. (See *Drury, Illustr. vol.* 2. *t.* 33.)

In his earlier works, Fabricius erroneously gave the Cape of Good Hope as the locality for this very common Chinese species.

RHINA BARBIROSTRIS.

Plate 4. fig. 3.

GENUS.	RHINA, *Latreille*. Curculio p. *Donovan, &c.*
CH. SP.	Rh. longirostris; niger, rostro barbato, tibiis anticis tridentatis. Long. Corp. (rostr. excl.) lin. 11.
	Rh. with a long rostrum, its sides thickly clothed with hairs on the male, black, punctured, anterior tibiæ long and armed with three spines beneath. Length without the rostrum 11 lines.
SYN.	Curculio barbirostris, *Fabricius Ent. Syst.* 1. *p.* 2. 418.
	Rhina barbirostris, *Latreille Genera, vol.* 2. *p.* 269.

Donovan, misled by the Fabrician reference of this insect to the locality " In Indiis," introduced it into the present work. It is, however, a native of South America. Drury has figured the female of another species in his Illustrations, vol. 2. pl. 34. f. 2. from the Island of Johanna.

c

HIPPORHINUS VERRUCOSUS.

Plate 4. fig. 4.

DIVISION. ENTIMIDES, *Schonherr.*
GENUS. HIPPORHINUS, *Schonh.* Hipporhis, *Bilberg.* Curculio p. *Auct.*
CH. SP. H. elongato-ovatus, niger, æneo-micans, thorace confertim tuberculato, elytris seriato-tuberculatis apice singulatim verruca crassa auctis. Long. Corp. lin. 9.
H. elongate-ovate, of a black colour, slightly tinged with brass, thorax closely tubercled, elytra with three rows of elevated warts on each, with intermediate double rows of punctures, and a large tubercle at the tip of each. Length ¾ inch.
SYN. Curculio verrucosus, *Linn. Syst. Nat.* I. II. *p.* 618. *Fabricius Syst. Eleuth.* 2. 534. *Schonh. Syn. Ins. Curcul.* 1. *p.* 481. *Drury Illustr.* 1. *pt.* 32. *f.* 5.

Donovan correctly referred this insect to the Curculio verrucosus, but incorrectly gave it as an inhabitant of China. This species, as well as the numerous insects of which the genus Hipporhinus is composed, are inhabitants of the Cape of Good Hope or New Holland.

HYPOMECES SQUAMOSUS.

Plate 4. fig. 5. magn. nat.
Plate 5. idem magn. auct.
Plate 4. fig. 6. idem, var. β.

DIVISION. BRACHYDERIDES, *Schonherr.*
GENUS. HYPOMECES, *Sch.* Curculio p. *Auct.*
CH. SP. H. oblongus, niger, squamulis viridi-aureis undique tectus, thorace supra fere plano, ruguloso, longitudinaliter canaliculato elytris subtiliter punctato striatis. Long. Corp. lin. 7.
H. oblong black, entirely covered with golden green scales, thorax nearly flat above, and rugose with a longitudinal canal, elytra slightly punctate-striate. Length 7 lines.
VAR. β. Corpore squamulis cinerascentibus undique tecto. Entirely clothed with grey scales.
SYN. Curculio squamosus, *Fabricius Ent. Syst.* 1. *p.* 2. 452. *Syst. Eleuth.* 2. 510. *Oliv. Ent.* 5. *p.* 319. *t.* 5. *f.* 48. *a. b.* Schonherr Syn. *Ins. Curcul.* 2. *p.* 71.
VAR. β. *Schonherr* loc cit. Curculio pulverulentus, *Fabricius Syst. Eleuth.* 2. 510. *Donovan*, 1st. edit.
Curculio orientalis, *Oliv. Ent.* v. *p.* 321. *t.* 6. *f.* 66?

A small, but superb species, being totally covered with minute scales of an oblong form, and resplendent green colour, interspersed with changeable sparks of gold and

Hypomeces Squamosus

1. Lamia Rubus.   2. Lamia Reticulator.
3. Lamia punctator.

COLEOPTERA.

crimson, in various reflections of light. This scaly covering is not unlike that of Polydrusus argentatus found in England; but of a brilliance scarcely inferior to the gem-like spangles on the Entimus imperialis of Brazil.—Hypomeces squamosus is represented in its natural size, in the annexed plate; and, in justice to an insect of such uncommon beauty, an additional plate is given, to exhibit its appearance in the opaque microscope. —It is extremely common in China.

The insect represented in fig. 6 is entirely black when divested of its scaly covering. Fabricius considered it as specifically distinct from H. squamosus, which opinion was adopted by Donovan. It is, however, considered by Schonherr, and other recent writers, as a variety only of that species, the C. unicolor, Fabr. and C. rusticus, Weber, being also equally regarded as varieties.

CLEONIS PERLATUS.
Plate 4. fig. 7.

Division.	CLEONIDES, *Schonherr*.
Genus.	CLEONIS, *Schonherr*. Cleonus, *Dejean*. Curculio p. *Linn. &c.*
Ch. Sp.	Cleonis niger, subtus densè, supra tenuiter cinereo-albido tomentosus, thorace vittis sub-tribus albidis notato, elytris punctato-striatis, ventre tuberculis numerosis glabris atris notato. Long. Corp. lin. 6.
	C. black; beneath densely, above slightly clothed with greyish scales, thorax with three slight whitish stripes; elytra punctate-striate, abdomen marked with numerous black smooth tubercles. Length ½ inch.
Syn.	Curculio perlatus, *Fabricius Syst. El.* 11. p. 516.
	Lixus faunus, *Oliv. Ent.* V. p. 267. *t.* 24. *f.* 342.

LAMIA RUBUS.
Plate 6. fig. 1.

Tribe.	LONGICORNES, *Latr.* (Cerambyx *Linn.*)
Family.	LAMIIDÆ.
Genus.	LAMIA, *Fabricius*. Cerambyx p. *Linn, &c.*
Ch. Sp.	L. grisea, thorace utrinque spinoso bimaculato, elytris basi scabris, humeris apiceque mucronatis, albo-guttatis. Long. Corp. unc. 1. lin. 9.
	L. grey coloured, thorax on each side with a spine and with two oblong spots on the back, elytra rough at the base, with the shoulders and apex spined, spotted with white. Length 1¾ inch.
Syn.	Cerambyx Rubus, *Linn. Syst. Nat.* 1. II. 625. 21.
	Lamia Rubus, *Fabricius Ent. Syst.* 2. 290. *Oliv. Ent.* 67. *t.* 7. *f.* 57.

This is the largest species of this genus found in China. It is also very abundant in different parts of the East Indies. Some interesting observations upon its habits have been published by W. W. Saunders, Esq. F. L. S. in the first part of the Transactions of the Entomological Society of London.

LAMIA RETICULATOR.

Plate 6. fig. 2.

Ch. Sp.   L. nigra, thorace elytrisque fulvis, thorace nigro-lineato, elytris reticulatis, antennarum articulo 3^{tio} fasciculato. Long. Corp. lin. 13.

L. black, thorax and elytra rich golden brown, the former marked with black lines, and the latter with irregular black marks, antennæ with the third joint furnished with a whorl of hairs. Length 13 lines.

Syn.   Lamia reticulator, *Fabricius Ent. Syst.* 1. *p.* 2. 278. *no.* 44.   *Oliv. Ins.* 67. *tab.* 12. *f.* 85.

This is altogether a beautiful insect; but the singular structure of the antennæ deserves particular notice: it is entirely brown except the first articulation, which is black; the third has a large verticillated tuft of black hair at the summit; at the base of this articulation it has another tuft, but smaller; and a similar tuft, but still smaller, is situated on the two following articulations.

LAMIA PUNCTATOR.

Plate 6. fig. 3.

Ch. Sp.   L. atra, elytris albo punctatis, antennis longis articulis albo nigroque variis. Long. Corp. 1 unc. 3 lin.

L. black shining, the elytra with farinaceous white spots, the antennæ long, the joints varied with black and white. Length 1¼ inch.

Syn.   Lamia punctator, *Fabr. Sp. Ins.* 1. 221. 30.   *Syst. Eleuth.* 2. 298.   *Oliv. Ent.* 4. 69. *t.* 8. *f.* 50. *a. b.*   *Schonh. Syn. Ins.* 3. *p.* 386.
Cerambyx Chinensis, *Forster Cent. Ins.* 39.
Cimex farinosus, *Drury, Edit.* 1ma *vol.* 2. *pl.* 31. *f.* 4.   *Donovan, Edit.* 1^{ina} (nec *Linn. Syst. Nat.* 1. 2. 626.)

Donovan states that among some Chinese drawings of the late Mr. Bradshaw, he observed one on which the metamorphosis of this insect was delineated. The larva was partly concealed in the hollow of a piece of decayed wood; it was of a whitish colour, with the head and tail black, as described by Fabricius. The true Cerambyx farinosus of Linnæus, with which this insect was confounded by Donovan, is an inhabitant of South America.

NATURAL HISTORY OF THE INSECTS OF CHINA

1. Buprestis Vittata. 2. Buprestis Ocellata.

BUPRESTIS (CHRYSOCHROA) VITTATA.

Plate 7. fig. 1.

TRIBE.	STERNOXI, *Latreille*.
GENUS.	BUPRESTIS, *Linnæus*.
SUB-GEN.	CHRYSOCHROA, *Carcel et Delap*.
CH. SP.	B. aureo-viridis, elytris bidentatis, punctatis; lineis quatuor elevatis, vittaque lata aurea. Long. Corp. 1 unc. 5 lin.
	B. golden green, elytra with two apical teeth, punctured, with four elevated lines and a broad golden stripe. Length 1 inch 5 lines.
SYN.	Buprestis Vittata, *Fabricius Syst. Eleuth.* 2. *p.* 187. *Oliv. Ent.* 2. 32. *p.* 9. *t.* 3. *f.* 17. *a—d. Herbst. Col.* 9. *tab.* 138. *f.* 4.
	Buprestis ignita, *Herbst. Archiv. t.* 28. *a. f.* 3.

The family Buprestidæ is one of the most extensive and brilliant tribes of coleopterous insects. Brasil and New Holland produce some gigantic species, but none more beautiful than those of India. We need adduce no other proof of this, than Buprestis chrysis, sternicornis, attenuata, ocellata, and vittata. These wrought into various devices and trinkets decorate the dresses of the natives in many parts of India. The Buprestis vittata in particular is much admired among them. It is, we believe, entirely peculiar to China, where it is found in vast abundance, and distributed from thence at a low price among the other Indians. The Chinese, who always profit by the curiosity of Europeans, collect vast quantities of this Buprestis, and other gay insects, in the interior of the country, and traffic with them.

The Buprestis ignita of Linnæus, with which the present species has been partially confused, has not the brilliance of colours that so eminently distinguishes B. vittata, but in form and size it agrees with it. The only figure of that species is given by Olivier, from a specimen formerly in the cabinet of Gigot d'Orcy, of Paris. The specimen in the cabinet of Sir J. Banks, referred to by Fabricius as B. vittata, agrees with Sulzer's figure of that species, as well as the specimen represented here, so that the reference by Fabricius of Sulzer's figure to B. ignita is incorrect.

Fabricius has given as a part of the specific distinction of these insects, that B. ignita has *three spines* at the end of each wing case, or elytron, and B. vittata no more than *two*. This may form a sufficient characteristic in those species; but we must remark, that it is not so in Buprestis ocellata. We have two specimens that have two spines at the end of each elytron, and another with three, as Fabricius has described it. We also find several insects nearly allied to B. vittata, the stripe of gold on each side excepted; one of these has six teeth, another four teeth, and a third only two.

Donovan observes that the Buprestides are supposed, for the most part, to undergo their

COLEOPTERA.

transformations in the water, or marshy ground. This opinion cannot, however, be adopted, as it is now well ascertained that they reside in the early stages of their existence in timber. The Chinese plant represented is the Canna Indica, or Indian flowering Reed.

BUPRESTIS (CHRYSOCHROA) OCELLATA.
Plate 7. fig. 2. upper, and 2ª under side.

Ch. Sp. B. viridi-nitens, elytris lineis tribus elevatis, macula ad basin alteraque apicali aureis, ocello magno flavo. Long. Corp. 1 unc. 4 lin.

B. shining-green, elytra with three elevated lines, a large yellow round spot in the middle of each, having a golden red spot above, and another behind it. Length $1\frac{1}{3}$ inch.

Syn. Buprestis Ocellata, *Fabricius Syst, Eleuth.* 2. p. 193. no. 38. *Oliv. Ent.* 2. p. 27. t. 1. f. 3. a. b. *Herbst. Col.* IX. p. 70. t. 144. f. 1. *De Geer Ins.* 7. 633. tab. 47. f. 12.

The Buprestis ocellata is very rare. Olivier says it is from *Chandernagore* in the East Indies. Mr. Drury possessed an extraordinary variety of this insect from China, in which the two spots united at the suture so as to form only one large spot on the back when the wing cases are closed.

These spots are strikingly characteristic of this species. They are situated in the centre of each elytron; are somewhat pellucid, and in fine specimens are cream colour, surrounded with a crimson circle. These spots are sometimes brown; probably they become so after the insect dies.

Our figures represent this insect with expanded wings; one of those is designed to exhibit the beautiful appearance of the under surface, particularly the effulgent abdomen and purple colour of the interior part of the elytra.

MYLABRIS CICHORII.
Plate 8. fig. 1. and 1ª.

Tribe. Trachelides, *Latreille.*
Family. Meloidæ.
Genus. Mylabris, *Fabricius.* Meloe p. *Linn.*
Ch. Sp. M. nigra elytris maculâ basali fasciisque duabus latis undatis fulvis. Long. Corp. 9—12 lines.

M. black, each of the elytra with a round basal spot, and two broad irregular bands of a fulvous colour. Length from $\frac{2}{3}$ths to 1 inch.

Syn. Meloe cichorii, *Linn. Syst. Nat.* 1. II. p. 680. 5. *Fabr. Syst. Eleuth.* 11. p. 81. 2. *Schonh. Syn. Ins.* vol. 3. p. 31. *Billb. Monagr. Mylabr.* p. 11. 4. t. 1. f. 8.

1. Mylabris Cichorii. 2. Sagra Splendida

中国昆虫志

COLEOPTERA.

This insect is very common in China and some other parts of the East Indies. The small specimen (fig. 1. a.) is rare, and is, probably, the male. According to Olivier, " the Cantharides of the ancients, and those of the Chinese, are not the same as ours. The Chinese employ the *Mylabris Cichorii*, and it appears from *Dioscorides Mat. Med. Lib. 2. Cap.* 65, that the ancient Cantharides were the same as those now used by the Chinese." " The most efficacious sort of Cantharides," says Dioscorides, " are of many colours, having yellow transverse bands; the body oblong, big, and fat; those of only one colour are without strength." The description Dioscorides has given does not agree with our species of Cantharides, as they are of a fine green colour, but is more applicable to the *Mylabre de la Cichorei*, which is very common in the country where Dioscorides lived. *Olivier's Entomologie, ou Hist. Nat. des Insectes. Vol. I. Introd.*

By the term Cantharides, in an European *Pharmacopœia*, we understand the Meloe vesicatorius* of Linnæus, an insect whose medicinal properties are very generally known.† The Cantharides of the ancients can scarcely be ascertained; it was a term indiscriminately applied to several kinds of insects, and too often without regard to their physical virtues. Pliny speaks of the Cantharis as a small beetle that eats and consumes corn; and of another that breeds in the tops of ashes and wild olives, and shines like gold. The ancients were certainly well acquainted with our common sort, though it is confounded with others in a general appellation.‡ *Hippocrates, Galen, Pliny, Matthiolus*, and other physical writers of antiquity, treat of the medicinal uses of Cantharides; but it is not clear that they alluded to only one species: indeed Dioscorides also mentions those of only one colour as being employed as vesicants. The ancients often confounded the term Scarabæus with Cantharis; but whether because they knew that the common kinds of Scarabæi produce the same effects as the Cantharis, is uncertain.—The *Scarabæus auratus*, and *Melolontha*, several *Coccinellæ, Cimex nigro-lineatus, &c. &c.* have a place in the *Materia Medica* as *Cantharides*.

---

* Geoffroy calls this a Cantharis. The Linnæan Cantharis is a distinct genus. (Telephorus, *Latreille*.)

† Applied externally to raise blisters. It is a violent poison taken inwardly, except in small portions.

‡ The common sort has been called *Musca Hispanica* by some Latin authors, and hence Spanish fly by Boyle.

SAGRA SPLENDIDA.

Plate 8. fig. 2. ♂ fig. 2ᵃ. ♀ ?

Tribe.	Eupoda, *Latreille.*
Family.	Sagridæ.
Genus.	Sagra, *Fabricius.* Tenebrio p. *Linn, &c.*
Ch. Sp.	S. cyaneo-purpurea, femoribus posticis dentatis, tibiis apice sinuatis. Long. Corp. lin. 9.
	S. glowing purple, changeable to yellow or green, posterior femora dentate, tibiæ with a deep notch at the tips. Length 9 lines.
Syn.	Sagra splendida, *Weber Obs. Ent. p.* 61. *Fabricius Syst. Eleuth.* 2. *p.* 27.
	Sagra femorata, *Donovan,* 1*st edit. in tab.*
	Tenebrio femoratus, *Donovan.* 1*st edit. in text.*

Donovan correctly observed, that the insect here figured differed from the figures referred to by Fabricius under Sagra femorata, the only described species at the period of the publication of the first edition of this work. Weber, however, subsequently published a monograph upon the genus, and this Chinese species appears to be identical with his Sagra splendida. The smaller figure appears also to be identical with the Sagra purpurea of Weber, which Fabricius was inclined to regard as the male of the former, although he gave it as distinct with the character:—" Sagra purpurea, nitida, femoribus posticis unidentatis tibiis integris." It appears to me, however, that it is merely the female of S. splendida.

Order. ORTHOPTERA. *Olivier.*

MANTIS (SCHIZOCEPHALA) BICORNIS.

Plate 9. fig. 1.

Tribe.	Cursoria, *Latreille.*
Family.	Mantidæ.
Genus.	Mantis, *Linn.*
Sub-Gen.	Schizocephala, *Serville.*
Ch. Sp.	M. thorace filiformi lævi, testaceo, oculis oblongis porrectis acuminato spinosis, elytris alis brevioribus. Long. Corp. 4¼ unc.
	M. with the thorax filiform smooth and pale buffish coloured, eyes oblong, porrected, and produced into a sharp point, elytra shorter than the wings. Length 4¼ inches.
Syn.	Mantis bicornis, *Linn. Syst. Nat.* I. 11. *p.* 691.
	Mantis oculata, *Fabricius Ent. Syst. t.* 2. *p.* 19.
	Schizocephala stricta, *Serville Revis. Orthopt. p.* 29 ?
	La Mante Chinoise étroite cornue. *Stoll. Repres. des Mantes.*

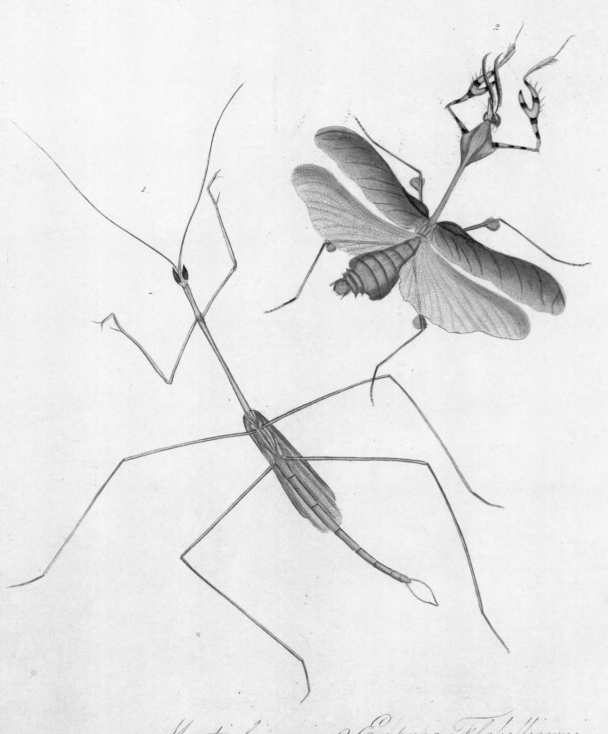

1. Mantis bicornis. 2. Empusa Flabellicornis.

ORTHOPTERA.

Two figures of insects, very much resembling our species, are given in the work of Stoll, &c.; one kind he calls *La Mante étroitement cornue*, the other *La Mante Chinoise étroite cornue*. The first is from the coast of Coromandel and Tranquebar, the other, as its name implies, is a Chinese insect. Donovan states that he could not discover any material difference between these figures and the specimen here figured, and was inclined to consider them altogether as one species.

It is a considerable disadvantage to the works of Stoll, as well as to the naturalist who consults them, that no scientific names, or definitions, are given to the figures of many rare insects included amongst them, hence they have been but rarely referred to by Fabricius.

The Mantis oculata of Fabricius is an African insect, and was described from the collection of the Right Hon. Sir J. Banks, Bart., Donovan compared his Chinese specimen with it, and found it precisely the same species.

MANTIS (EMPUSA) FLABELLICORNIS.

Plate 9. fig. 2.

SUB-GEN. EMPUSA, *Latreille*. Gongylus, *Thunberg*.

CH. SP. M. thoracis parte anticâ dilatatâ membranaceâ; femoribus anticis spinâ, reliquis lobo terminatis, antennis pectinatis. Long. Corp. $2\frac{1}{2}$ unc.

M. with the front of the thorax furnished with a large membrane, the two anterior thighs terminated by a spine, and the four posterior by a rounded membranous lobe, antennæ pectinated. Length $2\frac{1}{2}$ inches.

SYN. Mantis flabellicornis, *Fabricius, Ent. Syst.* II. *p.* 16. *no.* 16. *Serville Revis. Orthopt. p.* 21. (Empusa fl.)

This Mantis is described by Fabricius only. Stoll has given the figure of an insect not unlike it in his publication; and we have seen a specimen similar to it, which was found by Professor Pallas near the Caspian sea. It is allied to Mantis Gongylodes,[*] a native of Africa and Asia, but bears a closer affinity to Mantis Pauperata[†] from Java, Molucca, and perhaps other islands in the Indian sea.

Fabricius enumerates fifty-one species of this genus in his last system; a considerable portion of these are from Asia: had he included the America and New Holland species, his genus would have been far more comprehensive. Few naturalists have had the opportunity of observing the manners of these creatures in distant countries; nor can we always rely on the information those few have given. Of the European species we can

---

[*] Serville, indeed, considers it possible that M. flabellicornis may be the male of M. gongylodes.
[†] Figured by Stoll under the name of *La Mante Goutteuse Brune?*

ORTHOPTERA.

speak with more precision, because some indefatigable naturalists have attended minutely to them; Roesel in particular has treated at considerable length on the manners of the Mantis religiosa of Linnæus.

Descriptions can only convey an imperfect idea of the extraordinary appearance of many creatures included in the genera Mantis and Locusta. Among them are found species that bear a similitude to the usual forms of other insects; but, from these we almost imperceptibly descend to others, bearing as strong a similitude to the vegetable part of creation; seeming as if Nature designed them to unite the appearance of a vegetable with the vital functions of an animal, to preserve them from the ravages of voracious creatures, or to connect that chain of progressive and universal being, which

"The great directing MIND of ALL ordains."

Many of these creatures assume so exactly the appearance of the leaves of different trees, that they furnish the entomologist with unerring specific distinctions; thus we have *L. citrifolia, laurifolia, myrtifolia, oleifolia, graminifolia*, and others, equally expressive of their resemblance in form, and colours, to the leaves of those respective plants. Travellers, in countries that produce these creatures, have been struck with the phenomenon, as it must appear, of animated vegetable substances; for the manners of the Mantis, in addition to its structure, are very likely to impose on the senses of the uninformed. They often remain on the trees for hours without motion, then suddenly spring into the air, and, when they settle, again appear lifeless. These are only stratagems to deceive the more cautious insects which they feed upon; but some travellers who have observed them, have declared they saw the leaves of those trees become living creatures, and take flight.

M. Merian informs us of a similar opinion among the Indians, who believed these insects grew like leaves on the trees, and when they were mature, loosened themselves and crawled, or flew away. But we find in the more pretending works of Piso similar absurdities.

"Those little animals," says that author, "change into a green and tender plant, which is of two hands breadth. The feet are fixed into the ground first; from these, when necessary humidity is attracted, roots grow out, and strike into the ground; thus they change by degrees, and in a short time become a perfect plant. Sometimes only the lower part takes the nature and form of a plant, while the upper part remains as before, living and moveable: after some time the animal is gradually converted into a plant. In this Nature seems to operate in a circle, by a continual retrograde motion."*

---

* Donovan quoted, in a note, Ovid's account of the Transformation of Phaeton's Sisters into trees. "Luna quater junctis implerat cornibus orbem," &c. which he seems to think had its origin in some such idea as this.

ORTHOPTERA.

Roesel treats this account with more than merited severity; not because he could contradict the relation of Piso, but, because he had never observed the same circumstance attend the Wandering Leaf, or Mantis Oratoria, in Europe;* although he afterwards describes even the first symptom of the transformation as related by Piso. When he speaks of the death of the European species, his words are, "As their dissolution approaches, their green eyes become brown, and they unavoidably lose their sight: they remain a long while on the same spot, till at last they fall quite exhausted and powerless, as if asleep." As to the change after they remained long on the ground, such as sending forth fibres, roots, and stems, from the body of the insect, it is only astonishing such a well-informed naturalist should have deemed it matter of surprise. Could he be ignorant of the many instances that occur, of animal substances producing plants?† or was he not informed that the pupa which commonly sends forth a bee, a wasp, or cicada, has sometimes become the nidus of a plant, thrown up stems from the fore part of the head, and changed in every respect into a vegetable, though still retaining the shell and exterior appearance of the parent insect at the root?‡ We own at first sight with Roesel that the account of Piso seems "an inattentive and confounded observation," but that an insect may strike root into the earth, and, from the co-operation of heat and moisture, congenial to vegetation, produce a plant of the *cryptogamic* kind, cannot be disputed. We have seen species of *clavaria* both of the undivided and branched kinds, four times larger than the insect from which they sprang; and can we then deny that the insect mentioned by Piso might not produce a plant of a proportionate magnitude? In short, we are not sufficiently acquainted with the productions of Brazil to contradict any of his assertions, concerning this transformation. Piso does not say of what kind this vegetable was; it must surely be of the fungi kind: reasoning then from analogy, it might be an unknown species of *clavaria* with numerous and spreading branches; and, finally, the colour of

---

* Among the annotations on the last edition of Roesel's *Insecten Belustigung* we find one relating to this part of the works of Piso. "*Der seel Her geheime Rath Trew, &c.* Couns. Trew assures Mr. Roesel that *Piso* not only very often gave out the credible observations of others as his own, but himself believed the most incredible relations, and pretended to be an eye witness thereof." We quote this in justice to the remarks of Roesel. *Note in page 10, section Das Wandlende Blat.*

† Such as Mucor crustaceus, &c.

‡ Specimens of these vegetated animals are frequently brought from the West Indies; we have one of the cicada from the pupa, as well as others produced from wasps and bees in the perfect or winged state. Mr. Drury had a beetle in the perfect state, from every part of which small stalks and fibres have sprouted forth; they are very different from the tufts of hair that are observed on a few coleopterous insects, such as the Buprestis fascicularis, of the Cape of Good Hope, and are certainly a vegetable production.

his plant, on which authors lay much stress, might be green, though a colour not so predominant in that tribe of vegetables as some others.*

The largest and most interesting of the Indian species of Mantis is found in the isle of Amboyna. Stoll contradicts the account of *Renard*,† who says these creatures are sometimes thirteen inches in length; but we have a specimen almost of that size.‡ It is related by Renard, and others, that the larger kinds of Mantes go in vast troops, cross hills, rivers, and other obstacles that oppose their march, when they are in quest of food. If they subsisted entirely on vegetables, a troop of these voracious creatures would desolate the land in their excursions; but they prefer insects, and clear the earth of myriads that infest it: if these become scarce from their ravages, they fight and devour one another. When they attack the plants, they do great mischief. It is said of some Locusts and Mantes that the plants they bite wither, and appear as if scorched with fire: we have not heard of this pestilential property in any of the larger species of Mantes.

Of the smaller kinds, the Mantis Oratoria is the most widely diffused, being found in Africa and Asia as well as in all the warmer parts of Europe. These creatures are esteemed sacred by the vulgar in many countries, from their devout or supplicating posture. The Africans worship them; and their trivial names in many European languages imply a superstitious respect for them.§

England produces no species of this tribe. The entomologists in this country must consequently rely on the accounts of those, who have observed them in other parts of the world. We shall select a few remarks from Roesel's extensive description of Mantis Oratoria and Gongyloides, because, if we may presume from the analogy they bear in form to Mantis Flabellicornis, the history of one will clearly elucidate that of the other.

Roesel says, some of the Mantes are local in Germany; they are found chiefly in the vintages at Moedting in *Moravia*, where they are called *Weinhandel*.‖ The males die in October, the females soon after.¶ The young brood are preserved in the egg state, in a kind of oblong bag, of a thick spongy substance; this bag is imbricated on the outside;

---

* These arguments of Donovan, although sufficiently ingenious, prove only the accidental possibility of the insect producing plants, and not the transformation which Piso believed to be the ordinary nature of the creature. (I. O. W.)

† Poissons des Molucques par M. Renard, Amsterl. 1754.

‡ Donovan here evidently alludes to some of the Phasmidæ.

§ Louva Dios by the Portuguese. Prie Dieu by the French.

‖ Probably a provincial term for a dealer in wine.

¶ Goetz, in his Beytrage, observes, that they live sometimes ten years.

ORTHOPTERA.

it is fastened lengthwise to the branch of some plant.* As the eggs ripen they are protruded through the thick substance of the bag, and the larva, which are about half an inch in length, burst from them. Roesel, wishing to observe the gradual progress of these creatures, to the winged state, placed the bag containing the eggs in a large glass, which he closed, to prevent their escape. From the time they were first hatched they exhibited marks of a savage disposition. He put different sorts of plants into the glass, but they refused them, preying on one another: this determined him to supply them with other insects to eat: he put *ants* into the glass to them, but they then betrayed as much cowardice as they had barbarity before; for the instant the Mantes saw the ants they tried to escape in every direction. By this Roesel found the ants were the greatest persecutors of the Mantes. He next gave them some of the common musca (house flies), which they seized with eagerness in their fore claws, and tore in pieces: but, though these creatures seemed very fond of the flies, they continued to destroy one another through savage wantonness. Despairing at last, from their daily decrease, of rearing any to the winged state, he separated them into small parcels in different glasses; but here, as before, the strongest of each community destroyed the rest.

Another time, he received several pairs of Mantes in the winged state; profiting by his former observation, he put each pair [a male and female] into a separate glass, but they still shewed signs of an eternal enmity towards one another, which neither sex nor age could soften; for the instant they were in sight of each other, they threw up their heads, brandished their fore legs, and waited the attack: they did not remain long in this posture, for the boldest throwing open its wings, with the velocity of lightning, rushed at the other, and often tore it in pieces with the crockets and spines of the fore claws. Roesel compares the attack of these creatures to that of two hussars; for they dexterously guard and cut with the edge of the fore claws, as those soldiers do with their sabres, and sometimes at a stroke one cleaves the other through, or severs its head from the thorax. After this the conqueror devours his vanquished antagonist.†

We learn from Roesel also, the manner in which this creature takes its prey, in which respect we find it agrees with what is related of the extra European species. The patience of this Mantis is remarkable, and the posture to which superstition has attributed devotion, is no other than the means it uses to catch it. When it has fixed its eyes on an

---

* To that of the vine by Mantis Oratoria.

† The Chinese take advantage of these savage propensities, and keep these pugnacious insects in little bamboo cages, training them to fight for prizes, as cocks are fought in this country. This custom is so common, that, according to Mr. Barrow, (Travels in China) " during the summer months, scarcely a boy is to be seen without his cage of these insects." (J. O. W.)

ORTHOPTERA.

insect, it very rarely loses sight of it, though it may cost some hours to take. If it sees the insect a little beyond its reach, over its head, it slowly erects its long thorax, by means of the moveable membranes that connect it to the body at the base; then, resting on the four posterior legs, it gradually raises the anterior pair also; if this brings it near enough to the insect, it throws open the last joint, or crocket part, and snaps it between the spines, that are set in rows on the second joint. If it is unsuccessful it does not retract its arms, but holds them stretched out, and waits again till the insect is within its reach, when it springs up and seizes it. This is the uncommon posture before alluded to. Should the insect go far from the spot, it flies or crawls after it, slowly on the ground like a cat, and when the insect stops, erects itself as before. They have a small black pupil or sight which moves in all directions within the parts we usually term the eyes, so that it can see its prey in any direction without having occasion to disturb it by turning its head.

The most prevalent colour of this tribe of insects is fine green, but many of these fade or become brown after the insect dies: some are finely decorated with a variety of vivid hues; the most beautiful of these that we have seen are from the Moluccas.

TRUXALIS CHINENSIS.

Plate 10. fig. 1.

SECTION. SALTATORIA, *Latreille.* (Gryllus, *Linn.*)
FAMILY. LOCUSTIDÆ. (Acridites, *Latreille.*)
GENUS. TRUXALIS, *Fabricius.* (Gryllus, Acrida, *Linnæus.*)
CH. SP. Tr. viridis, capite thoraceque vittis quatuor, elytrorum lineâ centrali sanguineis, alis albido hyalinis. Expans. alar. $5\frac{1}{4}$ unc.
Tr. green, with four longitudinal stripes on the head and thorax, and a central line along the tegmina pink, wings stained pale buff hyaline. Expanse of the wings $5\frac{1}{4}$ inches.
SYN. Truxalis Chinensis, *Westw.*
Gryllus nasutus, *Donovan, 1st Edition.*

Donovan considered this insect as a variety of the Linnæan Gryllus nasutus, which he states to be found in Africa, Asia, and the south of Europe; adding, its varieties are numerous; and in size and colour depend on the climate they breed in. Sulzer represents it with red wings: in the Chinese specimens these are tinged with green. As several species are thus evidently confounded together, I have separated that here figured under the specific name of T. Chinensis.

2. Truxalis vittatus. 1. Truxalis Chinensis.

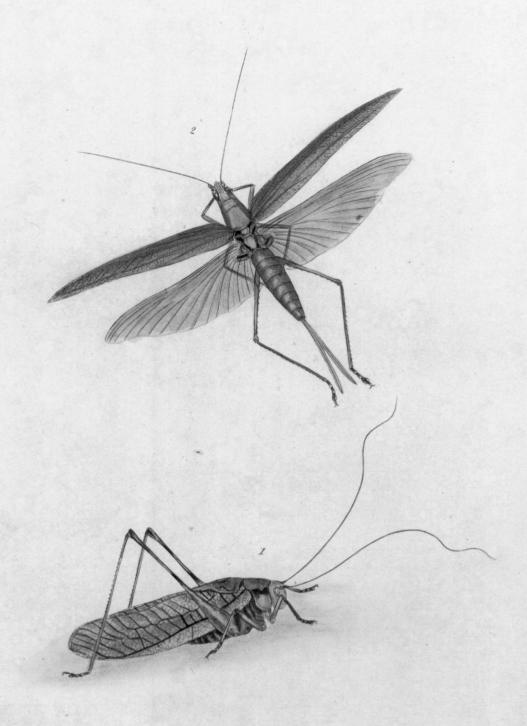

1. *Gryllus perspicillatus.*　2. *Gryllus acuminatus.*

ORTHOPTERA.

TRUXALIS (MESOPS) VITTATUS.

Plate 10. fig. 2.

Sub-Gen. Mesops, *Serville.*
Ch. Sp.   Tr. capite prominulo, testaceus, capite thorace femoribusque posticis vittâ laterali argenteâ. Long. Corp. 1⅓ inch.
         Tr. with the head slightly prominent; of a testaceous colour, with the head, thorax and posterior femora marked with a lateral silvery stripe. Length of the body 1⅓ inch.
Syn.   Truxalis vittatus, *Fabr. Ent. Syst.* 2. p. 27.

A single specimen of this insect, brought from China, and in the possession of Mr. Francillon, was employed as the original of this figure. From the form of the head it appears to belong to Serville's sub-genus, Mesops, but the wings extend beyond the body.

GRYLLUS (PHASGONURUS) PERSPICILLATUS.

Plate 11. fig. 1.

Family.   Gryllidæ. (Gryllus Tettigonia, *Linnæus.*)
Genus.   Gryllus, *Linnæus.* (Locusta, *Latreille.*)
Sub-Gen. Phasgonurus, (*Westw. Steph.* Locusta, *Serville.*)
Ch. Sp.   G. capite pallido, antennis fuscis, thorace virescenti postice rotundato, elytris concavis viridibus nervosis; basi ocello dorsali fenestrato. (Long. Corp. elytr. claus. 2¼ unc.
         G. with the head pale, antennæ brown, thorax green, rounded behind, elytra concave, green, nervose, with a dorsal fenestrated ocellus at the base. Length, with the wings closed, 2¼ inches.
Syn.   Locusta perspicillata, *Fabricius Ent. Syst.* 2. p. 36.

Donovan states that Fabricius erroneously describes this insect as a native of America, and that it is not figured elsewhere. Fabricius refers to Dr. Hunter's Museum, now belonging to the University of Glasgow.

ORTHOPTERA.

GRYLLUS (CONOCEPHALUS) ACUMINATUS?

Plate 11. fig. 2.

Sub-Gen. Conocephalus, *Leach, Serville.*
Ch. Sp. G. "thorace rotundato, vertice subulato, alis virescentibus." (*Linnæus loc. cit. subtus.*) (Expans. alar. fig. Donov. 4 unc.)
G. with the thorax rounded, forehead pointed, wings green. Expanse of the wings, according to Donovan's figure, 4 inches.
Syn. Gryllus acuminatus? *Linn. Syst. Nat.* 2. 696. *Fabr. Ent. Syst.* 2. p. 39. *Serville Revis. Orthopt.* p. 52. (Conocephalus ac.)

Donovan states that the insect here figured inhabits China and every other part of India. Linnæus, however, referring to Gronovius and Sloane's Jamaica, gives America as the locality of his G. acuminatus; whilst Fabricius says, "Habitat in omni India et in Europa Australi." In this confusion of locality, I have thought it best to mark the specific name of this figure with doubt.

LOCUSTA (RUTIDODERES) FLAVICORNIS.

Plate 12. fig. 1.

Family. Locustidæ, *Leach.* (Acridites, *Latreille.*)
Genus. Locusta. (Acrydium, *Latreille.*)
Sub-Gen. Rutidoderes, *Westw.* Acridium, *Serville.*
Ch. Sp. L. thorace subcarinato; viridis, elytris immaculatis, alis basi rufis, tibiis posticis sanguineis flavo-serratis. Expans. alar. 4 unc.
L. with the thorax somewhat keeled; green, with the elytra not spotted, wings at the base red, posterior tibiæ sanguineous with yellow teeth. Expanse of the wings 4 inches.
Syn. Gryllus flavicornis, *Fabricius Ent. Syst.* 2. p. 52.

1. Locusta flavicornis 2. Gryllotalpa Chinensis.

Locusta morbillosa

ORTHOPTERA.

GRYLLOTALPA CHINENSIS.

Plate 12. fig. 2.

FAMILY. ACHETIDÆ. (Gryllus Acheta, *Linn.*)
GENUS. GRYLLOTALPA, *Latreille*.
CH. SP. G. luteo fulvescens, tibiis anticis 4-dentatis. Long. Corp. 1 unc.
G. of a fulvous-clay colour, anterior tibiæ with four teeth. Length 1 inch.
SYN. G. Chinensis, *Westw.*
Gryllus Gryllotalpa, *Donovan*, 1st edition.

The genus of Mole-crickets, Gryllotalpa, comprises several distinct, although very closely allied species, the inhabitants of various countries, differing especially in the number of teeth of the anterior tibiæ and the neuration of the tegmina. Donovan regarded them as varieties only, observing that the species here figured "differs in no respect from the European species except in size and colour," the English mole-cricket "being twice as large and of more of a mouse colour."

LOCUSTA (PHYMATEA) MORBILLOSA.

Plate 13.

FAMILY. LOCUSTIDÆ.
GENUS. LOCUSTA.
SUB-GEN. PHYMATEA, *Westw.* Phymateus, *Thunberg, Serville*.
CH. SP. L. thorace quadrato, rubro, verrucoso; elytris fuscis, flavido punctatis; alis rufis, nigro punctatis. Expans. alar. 4¼ unc.
L. with the thorax squarish, red, warty; elytra bluish-brown, with pale yellow spots; wings red, with black spots. Expanse of the wings 4¼ inches.
SYN. Gryllus morbillosus, *Linn.* Syst. Nat. 2. 700. (e Cap. Bon Spei). *Fabricius Ent. Syst.* 2. *p.* 50. *Serville Revis. Orthopt. p.* 86. (Phymateus m.)

Donovan states that "the Gryllus morbillosus appears in the early edition of the *Systema Naturæ* and the works of Roesel as an Indian species, and that Mr. Drury assured him he had received it several times from China. Another sort is also found at the Cape of Good Hope, which is rather larger and deeper in colour than the Chinese variety."

If this be correct, it will, I apprehend, be necessary to consider these two *sorts* as

ORTHOPTERA.

distinct species, to retain the specific name morbillosus for the African species, and to give a new name to the Chinese species.

When this insect is at rest, the wings are folded and much of its beauty is concealed; but when these are expanded, its appearance is altogether magnificent. It has nothing of the shining and metallic splendour of the Coleoptera, for its colours are translucent, and assume their richest hues when they pass before the light. The elytra are purple, variegated with yellow; the wings of a glowing crimson, spotted with black; the abdomen is surrounded with alternate zones of black and yellow, and the legs are throughout of an elegant scarlet, inferior only in brightness to the coral red of the head and thorax. Upon the whole, this species is embellished with such a profusion of various and beautiful colours, that it may be considered as a most splendid example of the Linnæan Hemipterous order of insects. It is represented on the *Iris Chinensis* in a flying position.

This is not supposed to be a numerous species in China; on the contrary, it is probably uncommon. Several others of the locust are abundant in that country, and in seasons favourable to their increase do incredible mischief.* Both the Locusta tartarica and Locusta migratoria inhabit Tartary on the northern confines of China, from whence, at certain periods, they descend like an impetuous torrent over the neighbouring countries in quest of food, strip the earth of verdure, and scarcely leave the vestige of vegetation behind them. The Locusta migratoria, whose myriads are said to darken the face of heaven in their flights, sometimes direct their course westward, cross rivers, sea, and an immense extent of country, till they reach Europe; and though many are lost in these bold migrations, the survivors are in sufficient numbers to commit vast depredations. This species has been known to visit England,† but not in any abundance. In Little Tartary and the European provinces of Turkey, in Italy, and in Germany,‡ they do great mischief in these migrations. The Locusta flavicornis and

---

* " Famines sometimes happen in this part of the province; in some seasons inundations produced by torrents from the mountains, and as often the depredations of locusts, are causes of this disaster." (*Sir J. Staunton, Chap. on Tien-sing.*)

† The last appearance of this species in England was in 1748. Donovan had specimens of it from Smyrna, Germany, and China, and deemed it too common and general an inhabitant to merit a figure as a Chinese insect.

‡ Roesel speaks of this locust infesting the provinces of Wallachia, Moldavia, and Transylvania, in such immense numbers in the years 1747, 1748, and 1749, that an *Imperial and Royal Hungarian edict* was issued, with printed instructions for the best means of exterminating them. (*Der Heuschrecken-und Grillensammlung, &c. &c. vol. II. page* 193.)

NATURAL HISTORY OF THE INSECTS OF CHINA

nasuta are two other abundant species in China, and no doubt there are many other common kinds in that country we are at present unacquainted with. The locust is only detrimental when in immense numbers; for in China, as in other eastern countries, they are considered as an article of food, and regularly exposed for sale in the public markets.*

Order. HEMIPTERA. *Linnæus, Latreille.*

FULGORA CANDELARIA.

Plate 14.

SUB-ORDER.	HOMOPTERA, *Latreille*.
FAMILY.	FULGORIDÆ.
GENUS.	FULGORA, *Linnæus*.
CH. SP.	F. fronte rostratâ, adscendente; elytris viridibus luteo-maculatis; alis flavis apice nigris. Expans. alar. 3 unc.
	F. with the forehead produced into an ascending rostrum; elytra green, spotted with luteous; wings yellow buff, with the tips black. Expanse of the wings 3 inches.
SYN.	Fulgora candelaria, *Linn. Syst. Nat.* 2. 703. *Roesel Ins.* 2. 189. *t.* 20. *Sulzer Ins. t.* 10. *f.* 62. *Fab. Ent. Syst.* 4. *p.* 2. *Syst. Rh. p.* 2. *De Geer Ins.* 3. 197. 2. *Act. Holm.* 1746. *t.* 1. *f.* 5. 6.

The phenomena resulting from the properties and effects of light having engaged the attention of the earliest philosophers, we must conclude that phosphorical appearances, and those especially of animated bodies, could not fail to attract their particular notice. Indeed it is evident from the writings of the accurate observers of nature in remote ages, that they were acquainted with certain insects that have the property of shining in the night. These were known only by general terms, expressive of that property; yet it is probable that some of the Linnæan Lampyrides, which are abundant in the south of Europe, as well as in Asia and some parts of Africa, were the first of the illuminated

---

* Sir G. Staunton likewise speaks of "a large species of Gryllus" that is kept in cages for amusement in China, and was exposed for sale with other insects in the shops of Hai-ten. Neither the species of this, or the locusts noticed in the preceding note, are mentioned. (This remark evidently applies to the Mantidæ mentioned in the note to Empusa flabellicornis.)

HEMIPTERA.

insects known to them.* Some of the males, which are furnished with wings and are illuminated like the females, were striking objects of natural history, and could scarcely have escaped their notice. The Greeks included all shining insects under the name lampyris; and the Latins called them cicindela, noctiluca, luciola, lucernata, &c. Whether any of the *Fulgoræ* were known to the ancients is uncertain; probably they were not, the most remarkable species being peculiar to the warmest parts of America. Asia, once the seat of learning, does indeed produce a few species; but we have no account of these in ancient natural history.

The Fulgoræ seem to have been entirely unknown in Europe till the latter end of the seventeenth century, when two writers published descriptions and figures of *Fulgora Laternaria*; Madame Merian, of Holland, in her splendid work on the Metamorphoses of the Insects of Surinam, and Dr. Grew, of London, in his Rarities of Gresham College.

Reaumur† is the next author who described the *Fulgora Laternaria*, and after him Roesel, in his "Amusing History (or Recreation) of Insects."‡ This brings us to the period in which *Fulgora Candelaria*, our Chinese species, was first known in Europe; a circumstance of much importance to naturalists at that time, because the first-mentioned species was a solitary example of its singular genus. The transactions of the Stockholm academy includes the earliest figure and description of this extraordinary insect.

Roesel has given three figures and a description of it, and from his account we learn that it was known in England before he was acquainted with it. On its peculiar qualities he had been unable to derive any information concerning it, but his description is notwithstanding extremely prolix. We have selected the most interesting passage, because it clearly marks the progressive advancement of the knowledge of natural history in Europe so late as the middle of the present century.

"According to my promise," says Roesel,§ "I now produce the second sort of Lantern-carrier, which I never saw before, and of which I have never read in any work on insects. The scarcer, however, it may be, the more I am indebted to Mr. Beurer, apothecary of this place,‖ &c. for the permission he has granted me to draw and enrich my collection with it. Mr. Collinson has sent it to him from London, under the name

---

* The lampyris of *Pliny* is expressly the insect with a shining tail.
† Memoires pour servir a l'Histoire des Insectes. 1734.
‡ Insecten Belustigung.
§ Versprochener massen liefere ich nunmehr die zweite Sorte des Lanternen-Tragers, &c. Vol. i. pl. 30. Locust, page 189.
‖ Nurenberg.

HEMIPTERA.

Lanternaria Chinensis, for which reason I have called it the Asiatic or Chinese Lantern-carrier." Roesel being a respectable entomological writer of his time, we must infer that *Fulgora Candelaria* was extremely scarce in Europe when his plate and descriptions were published. The commercial concerns of Europeans with the Chinese having greatly increased since that period, has facilitated many inquiries concerning the natural productions of China; and amongst a variety of other insects that are now usually brought from that country, specimens of Fulgora Candelaria are extremely common. In China, few insects are found in greater abundance.

Having noticed the early history of this insect, we come to consider the peculiar properties of its singular genus; upon which the following observations were made by our author: " Among these we find the most astonishing that insects can possess, that of emanating light; not merely a momentary shining appearance, as is produced by many viscous substances, but a clear and constant resemblance to the element fire, and capable of diffusing light to surrounding objects, though totally destitute of every principle that can do mischief. To the unphilosophical mind it appears at first impossible, and it cannot fail to astonish the best informed; indeed, some readers might be inclined to doubt the veracity of travellers in foreign countries who have seen a vegetable* or an animal produce light, if our own country could not supply us with abundant analogous proofs of such phenomena. The presence of this animated phosphorus, if we may so express it, is observed on several insects that are natives of England; it is needless to enumerate them, because the most striking example must be recollected by every rural inhabitant or admirer of poetical simplicity:

> '——————— On every hedge
> The glow-worm lights his gem, and through the dark
> A moving radiance twinkles.'   THOMSON.

" The account which Madame Merian gave of the effect of the light produced by the *Fulgora Laternaria*† was greatly discredited, though Dr. Grew had related some sur-

---

* An instance of this occurs in the south of Europe. An account in the Philosophical Transactions relates of the Dictamnus Albus (Fraxinella), that " in the still evenings of dry seasons it emits an inflammable air, or gas, and flashes at the approach of a candle. There are certain instances of human creatures who have taken fire spontaneously, and been totally consumed."

† The account which Madame Merian has given of the light of the *Fulgora Laternaria* is so surprising, that it will certainly prove acceptable to many readers. It is indeed a digression from the account of *Fulgora Candelaria*, but will tend to prove, that insects of this genus emit a more vivid light than any of the illuminated kinds hitherto known.

" Once," says Madame Merian, " when the Indians brought me a number of these Lantern-carriers, I put

HEMIPTERA.

prising particulars of a specimen of it from Peru.* Her account has, however, been generally believed since the missionaries† in countries which produce those insects have confirmed her account. It is admitted that the *Chinese Fulgora* has an illuminated appearance in the night. 'The foreheads of many Fulgoræ (especially those found in China) emit a lively shining light in the night-time, which, according to some authors, is sufficient to read by.'—*Yeats.* ‡

"The light of the Fulgoræ is generally imagined to issue from the trunk, or elongated projection of the forehead; but Roesel offers a conjecture on the light of the Fulgora Laternaria, which, on further investigation, may enable naturalists to determine whether the light is entirely produced by an innate property of the trunk, or receives additional splendour from some external cause. He notices a white farinaceous substance on several parts of the wings and body as well as the trunk, which, he observes, looks like the decayed wood which shines at night. We mention this conjecture of Roesel, though the same occurred to us before we perused his observations. We have invariably found a similar white powder on other insects of this genus, but usually upon the trunk only. The remarks of Roesel were necessarily very limited, two species of the Fulgoræ only being then known. We possess twelve distinct species, and have made dissections and observations on several others; from all which we are inclined to imagine

---

them into a wooden box, without being aware of their shining at night; but one night, being awakened by an unusual noise, and much frightened, I jumped out of bed and ordered a light, not knowing whence this noise proceeded. We soon perceived that it originated in the box; we opened with some inquietude, but were still more alarmed after opening it, and letting it fall on the ground, for a *flame* appeared to issue from it, which seemed to receive additional lustre as often as another insect flew out of it. When we observed this some time we recovered from our terror, and admired the splendour of these little animals." *Dissert. de generatione et metamorphibus Insectorum Surinamensis.*

* Cucujus Peruvianus. "That which, beside the figure of the head, is most wonderful in this insect, is the shining property of the same part, whereby it looks in the night like a lantern, so that two or three of these fastened to a stick, or otherwise conveniently disposed of, will give sufficient light to those who travel or walk in the night." *Grew. Museum Regalis Societatis, p.* 158.

† Le ver-luisant. Ceux que nous voyons à la campagne dans les nuits d'été ne jettent qu'une foible lueur: mais ils y en a dans les Indes modernes qui répandent un éclat très-vif. Ce sont, pour ainsi dire, des phosphores animez. "Les Indiens," dit le savant auteur de la Théologie des Insectes, "ne se servoient autrefois dans leurs maisons, et dehors d'aucune autre lumière. Lorsqu'ils marchent de nuit, ils en attachent deux aux gros doigts du pied, et en portent un à la main. Ces insectes répandent une si grande clarté, que par leur moyen on peut lire, écrire, et faire dans une chambre toutes les autres choses nécessaires." *Lesser. Liv.* 2. *c.* 3. *rem.* 8. "Le trait rapporté par le P. du Tertre dans son *Histoire des Antilles*, auroit bien dû être cité, il dit avoir lû son bréviare à la clarté d'un de ces vers-luisans."

‡ Yeats. Institutions of Entomology.

Cicada Atrata.

HEMIPTERA.

that the white powder has a phosphoric appearance in the living insect, and increases the light when the end of the trunk is illuminated.

"One of the Fulgoræ of considerable magnitude, from the interior of India, enabled us to make many observations. The trunk is of the same form as that of the *Fulgora Candelaria*; the colour is a dark but beautiful purple; the apex scarlet, of a perfectly pellucid appearance, and still retains a reddish glare; the spots of white sprinkled on the purple colour of the trunk exhibit also a slight appearance of phosphoric matter. On the trunk of the Fulgora Candelaria these white spots are very conspicuous.

"Though the generic name Fulgora seems to imply some effulgent property in the insects that compose the genus, it is uncertain whether all possess that property. They are indeed furnished with a trunk, but it is smaller in proportion in several species than in F. Laternaria, Candelaria, Flammea, Phosphorea, and some others. It has not been determined whether any of the European Fulgoræ shine in the night-time." *

CICADA ATRATA

Plate 15.

Family. Cicadidæ.
Genus. Cicada, *Linnæus.* Tettigonia, *Fabricius.*
Ch. Sp. C. atra, alis albis basi nigris, venis testaceis. Expans. alar. 4¾ unc.
  C. black, with iridescent white wings black at the base, and with yellowish brown nerves. Expanse of the wing 4¾ inches.
Syn. Cicada atrata, *Fabr. Ent. Syst.* 4. *p.* 24. *no.* 28. *Syst. Rhyng. p.* 42.

Though the observations of Sir G. Staunton on the natural productions of China were necessarily very general, the study of insects appears to have engaged his particular attention; and on that account we must lament that untoward events precluded him from observing more minutely the peculiarities of some kinds, and the economical

---

* (Notwithstanding the observations of Donovan and the various authorities cited by him, it is certainly a matter of doubt, at the present day, whether the Fulgoræ in reality possess any luminous property. No recent author or traveller has noticed its existence in these insects, although it is related in every work of travels as exhibited by the Elateridæ and Lampyridæ. Moreover, the farinaceous matter noticed by Rösel exists in many other insects known to be not in the least degree luminous, and of which the woolly or waxen appendages in Lystra lanata, the Dorthesiæ, Cocci, &c. is but a modification or more extensive developement. M. Wesmael, of Brussels, has, however, just reasserted the luminous properties of the South American Fulgora on the authority of a friend who had witnessed it alive. *Ann. Soc. Ent. France*, 1837, *App.* J. O. W.)

HEMIPTERA.

purposes of others. We therefore peruse the following account of an unknown species of Cicada with particular regret, because it withholds information interesting to the naturalist, and from its air of novelty is likely to promote an erroneous opinion concerning that singular tribe of insects.

"The low and sometimes marshy country through which the river* passes, is favourable to the production of insects; and many of them were troublesome, some principally by their sting, and others by their constant stunning noise. The music emitted by *a species of Cicada* was not of the vocal kind; but produced by the motion of two flaps or lamellæ which cover the abdomen or belly of the insect. It is the signal of invitation from the male of *that species* to allure the female, which latter is quite unprovided with these organs of courtship."†

Again, when describing a town higher up the river, that author says, "The shops of Hai-tien, in addition to necessaries, abounded in toys and trifles, calculated to amuse the rich and idle of both sexes, even to cages containing insects, such as the *noisy Cicada*, and a large species of the Gryllus."‡

The reader may imagine from the first account, that the music of every other species of Cicada is of the vocal kind, or that it is peculiar to this Chinese insect to be furnished with lamellæ that cause a sound. The latter account confirms such conjecture, by alluding in a specific manner to the *noisy Cicada*, as to an insect described in a former part of the work. We must remark, that not only the males of the species mentioned by that author are furnished with those lamellæ, but the whole of that section of the Linnæan Cicadæ which Fabricius has called Tettigonia. The males of the species included in the other sections of that genus are certainly furnished with them also; though some of them are too minute to be observed without a glass. These lamellæ vary in size in different species; but the accounts we have of them from travellers in foreign countries, and naturalists both ancient and modern, prove that they all emit a certain sound to allure the female. As we are unable to ascertain the Chinese species Sir George mentions, neither figure nor description accompanying his account of it, we must therefore speak generally of the whole genus, and then confine our remarks to those species we are acquainted with from China. Among these are *C. splendidula, sanguinea*, and *atrata*. The latter, we believe, is the largest species of the Chinese Cicadæ known in Europe.

---

* Pei-ho.
† Chap. 3. Vol. II. octavo.
‡ Chap. 4. Vol. II.

HEMIPTERA.

Some species of this tribe were known to the ancients. With them it was the emblem of happiness and eternal youth;* and if we examine the legends of pagan mythology, we find they were deemed a race of creatures beloved by gods and men. The Athenians wore golden Cicadæ in their hair, to denote their national antiquity; or that like those creatures they were the *first born* of the earth; and the poets feigned that they partook of the perfection of their deities.† Anacreon depictures in glowing colours the uninterrupted felicity of this creature: his ode to the Cicada is appropriate to our inquiry.‡

In the infant state of music, men seem to have preferred the natural sounds of some animals to those of their uncouth instruments. We cannot otherwise account for the extravagant praise bestowed on the noise of this little creature. It is true, authors agree that the sounds of some kinds are exceeding loud and harmonious, and in the early ages of the world these might have a powerful influence on the human mind. It is related that the *ancient Locri*, a people of Greece, were so charmed with the song of the Cicada, that they erected a statue to its honour.§

The ancients had attentively observed the manners of its life, though they indulged

---

* Probably because it was supposed to live only a short time. The renewal of youth is illustrated by the story of the Tithonus transformed by Aurora into a Cicada.

† These pagan deities were without flesh or blood, and composed of aerial and watery humours. Such they imagined the moisture of the Cicada, and perhaps for that reason first assigned it a place among their demi-gods.

‡ Happy creature! what below
Can more happy live than thou?
Seated on thy leafy throne,
(Summer weaves the verdant crown,)
Sipping o'er the pearly lawn
The fragrant nectar of the dawn;
Little tales thou lov'st to sing,
Tales of mirth—an insect king:
Thine the treasures of the field,
All thy own the seasons yield;
Nature paints for thee the year,
Songster to the shepherds dear:
Innocent, of placid fame,
What of man can boast the same?

Thine the lavished voice of praise,
Harbinger of fruitful days;
Darling of the tuneful nine,
Phœbus is thy sire divine;
Phœbus to thy notes has giv'n
Music from the spheres of heav'n:
Happy most, as first of earth,
All thy hours are peace and mirth;
Cares nor pains to thee belong,
Thou alone art ever young;
Thine the pure *immortal* vein,
Blood nor flesh thy life sustain;
Rich in spirits—health thy feast,
Thou'rt a demi-god at least.
Green's Trans. Ode 43.

§ Some say, that once a certain player of *Locri*, contesting in the art of music with another, would have lost the victory, by the breaking of two strings of his instrument, but a *Cicada* flew to his aid, and resting on the broken instrument, sung so well, that the Locrian was declared victor. The Locrians erected a statue to the Cicada as a testimony of their gratitude. It represented the player with the insect on his instrument.

F

HEMIPTERA.

in many poetical fictions concerning it; and particularly when they affirmed that it subsisted on dew. They have told us that it lives among trees,* which circumstance discountenances the opinion of those moderns who imagine the grashoppers† were the Cicadæ of the ancients.

Neither were they ignorant that the males only were furnished with those instruments which externally appear to produce its sound, or the purpose for which that sound was emitted;‡ though it was reserved for more accurate naturalists to discover the complex organs by which it was caused and modulated. Aldrovandus, near two centuries ago, described the lamellæ, which he compares to the fruit of some herbs, called by modern botanists *Thlaspi*.§

Among later naturalists who have noticed the Cicadæ of foreign countries, are Merian,‖ Margravius,¶ &c. Merian says, its tune resembles the sound of a lyre which is heard at a distance; and that the Dutch in the plantations of Surinam (where they are very plentiful) call it the Lyre-player.** Margravius, in his natural history of Brazil, compares it to the sound of a vibrating wire: he says the tune begins with *Gir, guir*, and continues with *Sis, sis, sis*. One species is called Kakkerlak†† in the Indies, perhaps because the sound emitted by it may be likened to the pronunciation of that word. Mr.

---

* *Dr. Martyn* supposed this refers to the smaller branches in hedges, rather than to the lofty trees in forests: we cannot entirely coincide with that opinion.

† *Grashopper. Cicada.* They live almost every where in hot countries. *Lovel. Hist. Animal.* containing the summe of all authors ancient and modern, *p.* 274, *&c. &c.*

Cicada, a Sauterelle,ᵃ or, according to others, a *balm cricket*.—Non est quod vulgò, a *grashopper*, vocamus; sed insectum longè diversum, corpore et rotundiore et breviore, qui arbusculis insidet et sonum quadruplo majorum edit. a *grashopper*, rectè *locustum* reddideris, *Morl* ex *Ray*. *Ainsworth*.

‡ Xenarchus, an old Grecian play-writer, used to say jocosely, that "the Cicadæ were very happy because they had silent wives." Aristotle also knew the sexual difference of them; he mentions them as a delicious food: he preferred the males when young, but more so the females before she laid her eggs.

§ Thalaspi parvum Hieraciifolium, sive Lunariam luteam Monspel. et Leucoium luteum marinum. *Lobel. Stirpium Adversaria nova, p.* 74.—*Aldrov*.

‖ Merian. Insecta Surinamensia.

¶ Georgi Margravii rer. nat. Brasiliæ. *Lib.* 7. *p.* 257.

** De Lierman.

†† *Scopoli, Carn.* Yeats describes the Kakkerlak of the American islands as a species of *Blatta*, cock-roaches. Are there not two insects of that name?—one of them is, we believe, a *Blatta*. Indeed, Latreille has made use of the name Kakkerlak for a genus of Blattidæ.

ᵃ *Sauterelle*, sorte d'insecte. A locust or grashopper. *Boyer*.—*Cigale*, a flying insect. The Cicada of the ancients, unknown in England. *Boyer*.

HEMIPTERA.

Abbot, an accurate observer and collector of natural history in *North America*, has discovered four new species of Cicada, one of them nearly equal in size to our Cicada Atrata. This, he says, was found in great abundance one season in some swampy grounds near the Susquehanna river, and was remarkable on account of its loud noise, which at a little distance resembled the ringing of *horse-bells*.*

Some naturalists have supposed that the sound of the Cicada is caused by the flapping of the lamellæ against the abdomen; and others, that it is only a noise occasioned by the rustling of the segments of the body in the contractile motion of that part. Beckman† imagines it is caused by beating the body and legs against the wings: he has endeavoured to explain the meaning of ancient authors, and deduce its etymology from that circumstance.‡

Reaumur and Roesel have dissected several of the Cicadæ, and discovered that the lamellæ cannot have that free motion necessary to cause such a sound; but that it is produced by some internal organs of the insect, and only issues through the opening concealed under the lamellæ as through the mouth of a musical instrument.§

---

* Communicated by Mr. Abbot, in North America, to Mr. Francillon, in London.

† *Roes. Insecten Bellustigung.*—CHRISTIANI BECKMANNI, Bornensis, manuductionem ad latinam linguam: nec non de originibus latinæ linguæ, &c.

‡ It is the common opinion that the word *Cicada* has its origin from *quod cito cadat*, which, after a general interpretation, implies that the Cicadæ *soon vanish*, or are *short-lived*. Beckman maintains that this opinion is absurd, and proves that its name is derived from singing, because φ ἄδειν signifies a sound produced by the motion of a little skin; and that *ciccum* or *cicum* is a thin little skin of a pomegranate that parts the kernels.—Beckman not knowing the insect, or not imagining that the *little skin* was an appendage to the abdomen, concluded it must mean the transparent wings, and consequently that the sound was produced by beating them against the body: but this interpretation, if applied to the lamellæ instead of the wings, will directly prove the origin of its name, and knowledge of the ancients.

§ For the satisfaction of the curious reader, we detail the most interesting particulars concerning the organization of these parts from *Reaumur's Histoire des Insectes*, and *Roesel's Verschiedene auslaendische sorten von Cicaden, &c.*

The music of the Cicada is not caused by the motion of the *lamellæ* as some have supposed. Reaumur observes, that although the *lamellæ* have a kind of moveable hinge, they have also a stiff and pointed tooth, or spine, that prevents them from being lifted far back; and if strained are very liable to be broken.

From the anatomical description of Roesel, we find that, within the two hollows that are seen when the lamellæ are lifted up, two very smooth skins are visible; these are highly polished, of nearly a semicircular shape, and reflect prismatic colours: there is between these a hard brown projection, or corner which unites with another piece above them in a longitudinal direction, to the under part of the breast. This longitudinal piece divides a triangular red space or field into two parts, one on the right side, and the other on the left. Above these, in a transverse direction, are seen two small yellow skins; the lamellæ in their natural position conceal these organs because they fold exactly over them.

Reaumur, in the exterior appearance of these parts, could discover nothing that could lead to determine the

HEMIPTERA.

The suppositions of these authors seem well founded; we have examined many species that were unknown to them, and find the spine before mentioned so placed in many insects as to prevent the motion of the lamellæ. We have a specimen from America, which, in addition to the usual organs of sound, have two large hollow protuberances or drums; one on each side of the abdomen; and must, we imagine, produce a louder sound than any yet discovered; a species very similar to this is also brought from New Holland.

The proboscis of these insects is a hard or horny tube, in which a very acute slender sucking-pipe is concealed. The horny tube is not unlike a gimlet in form, and is used by those creatures to bore through the bark of trees, to extract the juices on which it feeds. Linnæus has named the species of one division, in his System, *Mannifera*, because they had been observed to fly among ash trees, bore many holes in them, and when the manna had oozed out return and carry it off.

With this proboscis they bore holes in the small twigs of the extreme branches of trees and deposit their eggs in them, sometimes to the amount of six or seven hundred. As each cell contains no more than from twelve to twenty eggs, it does great damage to the trees they frequent. Stoll says, " the common one,* which is found at Surinam in the coffee plantations, greatly injures those trees; the females depositing their eggs in

---

organs of the sound; and he was not satisfied that the slight motion of the lamellæ on these parts could produce the loud singing noise of the Cicada. He opened a few cicadæ on the back part of the body, so that the inner structure of the under side was displayed, and especially the parts connected to the curious organs he had discovered under the lamellæ. At last he discovered two large muscles, which at their point of union formed a space almost square, and were connected with the red triangular fields he had observed on the under side: as he concluded these formed a material part of the organs he wished to discover, he examined them attentively, and found that, by moving them backwards and forwards, he could make a cicada sing that had been dead many months. Although the sound was not strong, it tended to prove that he had discovered the instrument that produced it.—In another part he says, it is evident the sound is caused by the little skins connected to the muscles, because when they were rubbed with a bit of paper they emitted that kind of sound.

Roesel has discovered two little pieces of horny substance that are connected by a sort of fibre within the skins, in the body, and he supposes when this is in motion, it strikes against the before-mentioned thin skins, and produces a sound, by the same means as a hollow body, or drum, when struck with a stick: and also that this noise may be varied or modulated by a slight motion of the lamellæ, but cannot be produced without the assistance of the internal nerves and muscles connected with the organs first described.

Authors agree that the Cicadæ of hot countries emit the loudest sound. It appears from the papers of Mr. Smeethman (who resided a considerable time in Africa) published by Mr. Drury, that the sound of some kinds peculiar to that part of the world is so loud as to be heard at half a mile distance: and that the singing of one within doors silences a whole company.—The same attentive observer says, the open parts of the country are never without their music, some singing in the evening, and others only in the day.

* La Cigale Vieilleuse. Cicada Tibicen.

HEMIPTERA.

the young shoots, and in holes they bore with their sheath. They live on the juices of the trees."

M. Merian gives a figure and account of the metamorphosis of a cicada found in Surinam. She has mistaken the winged insect to be only the pupa of the *Fulgora Laternaria*, which is too absurd to deserve contradiction; in other respects her account is interesting, and particularly that part which relates to the pupa state, or chafer, as it is termed. "The pomegranate tree," says Merian, "so well known in all other countries, grows also in the fields of Surinam. On them I have found a species of chafer, which is naturally very lazy, and consequently very easy to be caught. It carries underneath the head a long trunk, with which it easily penetrates the flowers, in order to extract the honey from them. On the 20th of May, when they were laying quite quiet, the skin of the back burst open, and green flies, with transparent wings, issued from them. These flies are found in abundance in Surinam, and have such a rapid flight, that it took me many hours to catch one."

The pupa Donovan received from China with Cicada atrata very much resembles that figured by Merian. It has the long sucking trunk or proboscis; but the most formidable of its weapons seem to be the fore feet, which are thick, strong, and armed with spines or teeth; with these it may do more injury to the plants, by tearing off the tender shoots, than by wounding the trunk to extract the moisture.

The upper and under side of a male of Cicada atrata are represented, not only to illustrate our preceding remarks, but because Donovan believed no figure had been given of it by any author, unless *De Zweite Chineesche cicade* of Stoll (Pl. 20. fig. 118.) is intended for this insect.

The general appearance of both sexes of Cicada atrata is very similar, except that the female is furnished with a sheath, and the male with lamellæ. The sheath of the female is partly concealed within a valve at the extremity of the abdomen, and is only protruded when the creature lays her eggs. In the figure of the under surface of a male insect, exhibited in the annexed plate, the lamellæ are distinguished by two stars: the single star denotes the situation of the spine, mentioned by Roesel and Reaumur.

The Camphor-tree, *Laurus Camphora*, is represented in the plate. The tree which produces the useful drug *camphor* is very abundant in Japan and China. Sir G. Staunton says, it is the only species of the laurel genus growing in China, where it is a large and valuable timber tree, and is never cut up for the sake of the drug; but that substance is obtained by decocting the small branches, twigs, and leaves, and subliming the camphor in luted earthen vessels. A purer sort is brought from the island of Borneo and Japan, which is supposed to be a natural exudation from the tree when the bark is wounded. Sir G. Staunton says, the Camphor-tree is felled in those countries for the sole purpose of finding the drug in substance among the splinters.

HEMIPTERA.

CICADA SANGUINEA.

Plate 16. fig. 1.

Ch. Sp.   C. nigra, facie, thoracis maculis duabus, abdomineque sanguineis. Exp. alar. $2\frac{1}{6}$ unc.
C. black, with the face, two spots on the thorax, and the abdomen blood red. Expanse of the wings $2\frac{1}{6}$ inches.

Syn.   Cicada Sanguinea, *De Geer Ins.* 3. 221. *tab.* 34. *fig.* 17. *Donovan,* 1st edit.
Tettigonia Sanguinolenta, *Gozen Entomol. Beitr.* II. *p.* 150. *Fabricius Ent. Syst.* 4. *p.* 25. *Germar in Silberm. Rev. Ent.*

The specific name sanguinea of De Geer having the priority, is here retained in preference to that of Fabricius (Ent. Syst.), which is identical with that of another species of the Linnæan Cicadæ, as well as with that which, in his Systema Rhyngotorum, he has proposed in lieu of C. hæmatodes of his earlier works, which is however distinct from the Linnæan C. hæmatodes, and for which it will be necessary to employ another specific name.

CICADA AMBIGUA.

Plate 16. fig. 2.

Ch. Sp.   C. olivacea, elytris hyalinis, marginibus anticis testaceis. Expans. alar. $3\frac{1}{4}$ unc.
C. olive coloured, with the elytra clear, the anterior margin testaceous. Expanse of the wings $3\frac{1}{4}$ inches.

Syn.   Cicada ambigua, *Donov.* 1st edit.

Received by Mr. Drury from China.

LYSTRA LANATA.

Plate 16. fig. 3.

Family.   Fulgoridæ.
Genus.   Lystra, *Fabricius.*
Ch. Sp.   L. elytris apice nigris, punctis cæruleis, fronte lateribusque rubris, ano lanato. Expans. alar. $1\frac{3}{4}$ unc.
L. with the elytra black at the tips, spotted with blue, with the front and sides of the head red, abdomen woolly at the extremity. Expanse of the wings $1\frac{3}{4}$ inch.

Syn.   Cicada lanata, *Linn Syst. Nat.* 2. 711. *Fabr. Ent. Syst.* 4. 30. *Syst. Rhyng.* p. 56. *Sulzer Ins.* t. 9. f. 11. *Stoll Cicad.* t. 10. f. 49.

1. Cicada sanguinea.
2. Cicada ambigua.
3. Lystra lanata.
4. Cicada splendidula.
5. Cercopis abdominalis.
6. Jassus frontalis.

HEMIPTERA.

One of the most beautiful species of the Indian Cicadæ. The wing cases are black, elegantly reticulated, and spotted with bright blue. At the extremity of the abdomen it has a tuft of long and very delicate hairs, intermixed with others that are rather convoluted and of a coarser texture. The whole of this insect, but particularly between the abdomen and wings, is sometimes profusely covered with a fine powder of a snowy whiteness, similar to that observed on the Flata limbata in the imperfect state; hence we may conclude it is also one of those insects which furnish the white wax* so highly esteemed in China.

CICADA SPLENDIDULA.

Plate 16. fig. 4.

Ch. Sp. C. elytris fusco-aureis, femoribus anticis incrassato-dentatis rufis. Long. Corp. alis claus. ¾ unc.

C. with golden brown elytra, the anterior femora incrassated, toothed and red, thorax and scutellum varied with yellow and black. Length, with the wings closed, ¾ inch.

Syn. Cicada splendidula, *Fabricius Ent. Syst.* 4. *p.* 25. *Syst. Rhyng. p.* 42.

Figured from the unique specimen in the collection of Mr. Drury, described by Fabricius.

CERCOPIS ABDOMINALIS.

Plate 16. fig. 5.

Family. Cercopidæ.
Genus. Cercopis, *Fabricius*.
Ch. Sp. C. atra nitida, thorace immaculato, elytris basi fasciaque media flavescentibus; abdomine sanguineo. Long. Corp. alis clausis ⅘ unc.

C. black shining, thorax without spots, elytra with the base and a central fascia yellowish, abdomen sanguineous. Length, with the wings shut, ⅘ inch.

Syn. Cicada abdominalis, *Donovan, 1st Edit.*
Cercopis Heros? *Fabr. Syst. Rhyng. p.* 89.

---

* Vide Sir G. Staunton's *Hist. Emb. China.*

TETTIGONIA FRONTALIS.

Plate 16. fig. 6.

GENUS.   TETTIGONIA, *Latreille, Germar.*
CH. SP.   T. pallida, occipite thoraceque punctis quinque nigris, fronte puncto nigro inter oculos, elytris sanguineis. Epans. alar. fere 1 unc.

T. pale, with five black spots on the head and thorax, and one in front between the eyes, elytra red. Expanse of the wings nearly 1 inch.

SYN.   Cicada frontalis, *Donovan,* 1st Edit.
Cicada cæruleipennis? *Fab. Syst. Rh. p.* 73.

FLATA NIGRICORNIS.

Plate 17.

GENUS.   FLATA, *Fabricius.*
CH. SP.   F. exalbida, alis deflexis, elytris punctis marginis interioris antennisque nigris. Expans. alar. fere 2 unc.

F. whitish, with the wings deflexed, the elytra being spotted with black along the posterior margin, antennæ black. Expanse of the wings nearly 2 inches.

SYN.   Flata nigricornis, *Fabr. Syst. Rhyng. p.* 45.
Cicada limbata, var. *Fab. Sp. Ins.* 2. *p.* 322. *Donovan,* 1st Edit.

This singular insect, and the plant on which it is represented, have an equal claim to attention, both as objects of natural curiosity, and importance in domestic economy. The larva is an elegant and beautiful creature, and China is indebted to its labours for the fine white wax so much esteemed in the East Indies. The plant is not less interesting, as it produces the vegetable tallow, in general use throughout the Chinese empire.

The novelty of these productions could not fail attracting the notice of those learned Europeans who were first permitted to reside in China, and whose object was to promote sciences and arts, as well as the christian knowledge. Both the Wax-insect and Tallow-tree are spoken of in their writings as extraordinary and peculiar advantages to the country. Du Halde, especially in his splendid work l'Histoire de la Chine, treats largely on these productions, in the sections *Cire blanche d'Insectes et l'arbre qui porte le suif.* His relations are, perhaps, too prolix, but they are evidently the result of attentive

observation, and serve to illustrate the Natural History, and economical purposes of the subjects we are noticing.

The following is the account given by the author: " *De la Cire Blanche, faite par des insectes, et nommée Tchang pe la, c'est-à-dire, Cire blanche d'insectes.*—Ki *dit.* La Cire blanche dont il s'agit ici, n'est pas la même que la cire blanche des Abeilles. Ce sont de petits insectes qui la forment. Ces insectes succent le suc de l'espèce d'arbres nommée *Tong tçin*, et à la longue ils le changent en une sorte de graisse blanche, qu'ils attachent aux branches de l'arbre.

"Il y en a qui disent que c'est la fiente de ces insectes, qui s'attachant à l'arbre, forme cette Cire, mais ils se trompent. On la tire en raclant les branches dans la saison de l'Automne; on la fait fondre sur le feu, et l'ayant passée, on la verse dans l'eau froide où elle se fige, et se forme en pains. Quand on l'a rompue, on voit dans les morceaux brisez, des veines comme dans la pierre blanche ou congélation nommée Pe che cao; elle est polie et brillante: on la mêle avec de l'huile, et on en fait des chandelles. Elle est beaucoup supérieure à celles que font les Abeilles.

" *Chi tchin* dit. Ce n'est que sous la Dynastie des *Yuen* qu'on a commencé à connoître la cire formée par ces insectes. L'usage en est devenu fort commun, soit dans la médecine, soit pour faire des bougies. Il s'en trouve dans les Provinces de *Se tchuen* de *Hou quang*, de *Yunnan*, de *Fo kien*, de *Tche kiang*, de *Kiang nan*, et généralement dans tous les quartiers du Sud-Est. Celle qu'on ramasse dans les Provinces de *Se tchuen* et d'*Yunnan*, et dans les *territoires* de *Hen tcheou*, et de *Yung tcheou* est la meilleure.

" L'arbre qui porte cette cire, a les branches, et les feüilles semblables à celles du *Tong çin*. Il conserve sa verdure durant toutes les saisons: Il pousse des fleurs blanches en bouquets durant la cinquième Lune; il porte des fruits en bayes, gros comme le fruit du *Kin* rampant.

" Quand ils ne sont pas mûrs, ils sont de couleur verte; et ils deviennent noirâtres, lorsqu'ils mûrissent, au lieu que le fruit de *Tong çin* est rouge. Les insectes qui s'y attachent sont fort petits. Quand le soleil parcourt les quinze derniers dégrez des Gémeaux, ils se répandent en grimpant sur les branches de l'arbre; ils en tirent le suc, et jettent par la bouche une certaine bave, qui s'attachant aux branches encore tendres, se changent en une graisse blanche, laquelle se durcit, et prend la forme de cire. On diroit que c'est de la gelée blanche que le froid a durcie.

" Quand le soleil parcourt les quinze premiers dégrez du Signe de la Vierge, on fait la récolte de la Cire, en l'enlevant de dessus les branches. Si l'on diffère à la cuëillir que le Soleil ait entièrement parcouru ce Signe, il est difficile de la détacher, même en la raclant.

" Ces insectes sont blancs quand ils sont jeunes, et c'est alors qu'ils font leur cire.

HEMIPTERA.

Quands ils deviennent vieux, ils sont d'un châtain qui tire sur le noir. C'est alors que formant de petits pelotons, ils s'attachent aux branches de l'arbre. Ces pelotons sont au commencement de la grosseur d'un grain de mil: vers l'entrée du printemps ils commencement à grossir et à s'étendre. Ils sont attachez aux branches de l'arbre en forme de grappes, et à les voir, on diroit que l'arbre est chargé de fruits. Quand ils sont sur le point de mettre bas leurs œufs, ils font leur nid de même que les chenilles. Chacun de ce nids ou pelotons contient plusieurs centaines de petits œufs blancs.

" Dans le tems que le soleil parcourt la seconde moitié du Taureau on les cüeille, et les ayant enveloppez dans des feüilles de *Yo* (espèce de simple à larges feüilles); on les suspend à différens arbres. Après que le Soleil est sorti du Signe de Gémaux, ces pelotons s'ouvrent, et les œufs produisent des insectes, qui sortant les uns après les autres des feüilles dont ils sont enveloppez, montent sur l'arbre où ils font ensuite leur cire.

" On doit avoir soin d'entretenir le dessous de l'arbre toûjours propre, et de le garantir des fourmis qui mangent ces insectes. On voit deux autres arbres auxquels on peut attacher les insectes, et qui porteront également de la cire; l'un qui se nomme *Tien tchu*, et l'autre qui est un espèce d'arbre aquatique, dont les feüilles ressemblent assez à celles du Tilleul.

" *Qualitez et effects de cette cire.*—Elle est d'une nature qui n'est ni froide ni chaude, et qui n'a aucune qualité nuisible. Elle fait croître les chairs, elle arrête le sang, elle apaise les douleurs, elle rétablit les forces, elle unit les nerfs, et rejoint les os, prise en poudre dont on forme de pillules, elles fait mourir les vers qui causent la phtisie.

" *Tchi hen* dit. La Cire blanche est sous la dénomination du métal: ses esprits corroborent, fortifient, et sont propres à ramasser et à resserrer. C'est une drogue absolument nécessaire aux chirurgiens: elle a des effects admirables, quand on la fait entrer avec de la peau de *Ho hoang* dans la composition de l'onguent, qui fait renaître et croître les chairs." *Du Halde, Vol. IV. p.* 495, large Folio, 1735.

Sir G. Staunton, in his learned work, has also described the Wax insect; he found it at Turon Bay, in Cochin China, and has caused it to be represented in a vignette plate, with the following description. " Among other objects of natural curiosity, accident led to the observation of some swarms of uncommon insects busily employed upon small branches of a shrub, then neither in fruit nor flower, but in its general habit bearing somewhat the appearance of a privet. These insects, each not much exceeding the size of the domestic fly, were of a curious structure, having pectinated appendages rising in a curve, bending towards the head, not unlike the form of the tail feathers of the common fowl, but in the opposite direction. Every part of the insect was, in colour, of a perfect white, or at least completely covered with a white powder. The particular

HEMIPTERA.

stem frequented by those insects, was entirely whitened by a substance or powder of that colour, strewed upon it by them. The substance or powder was supposed to form the white wax of the East. This substance is asserted, on the spot, to have the property, by a particular manipulation, of giving in certain proportions, with vegetable oil, such solidity to the composition as to render the whole equally capable of being moulded into candles. The fact is ascertained, indeed, in some degree, by the simple experiment of dissolving one part of this wax in three parts of olive oil made hot. The whole, when cold, will coagulate into a mass, approaching to the firmness of bees' wax."

From the accurate description and figures of the latter author, it is evident, the creature that produces the white wax of China, is an imperfect insect, or technically speaking, the pupa of an insect, which, in its mature state, is furnished with wings. This is clearly the fact, for the rudiments of wings are visible in the figures alluded to.*

Stoll, in his work on *Cimices* and *Cicadæ*, gives a figure of this immature insect under the title of *De Waldraagster* (Nymphe) or *La Cigale Porte Laine*, fig. 144, together with the winged insect at fig. 145; and it is on this authority the latter is introduced in the annexed plate. There is, indeed, much similarity between the pupa and the imago, and some striking characteristics are common to both. They agree in the structure of the antennæ and proboscis, or sucking trunk; the abdomen of the winged insect is also loaded with a fine white powder, and is furnished at the extremity with a tuft of down and hairs, similar to that so eminently conspicuous in the pupa state. We have, however, observed the white powder and tuft on the abdomen of Lystra lanata, and have reason to imagine it also forms a white wax, similar to that of the present species.

Fabricius, in his Species Insectorum, described this insect as a variety of Cicada limbata, which is of a light green colour, with a red margin; that which Stoll has figured, and with which this agrees, is of a pale brown, with a black margin. These are the *species* and *variety* Fabricius describes, for the specimens referred to by Fabricius, in the collection of Sir Joseph Banks, agree precisely with our insects. Fabricius notes the *habitat* Africa. Stoll received the green specimen from the Island of Ceylon; the pale sort from Africa. The larva we have represented is from China; and the imago was brought from the East Indies, by the late Mr. Ellis.

---

* This may account for a passage in *Gordon's* description of China, where he says, " In the plains" of Houquang " are vast numbers of little *worms* that produce wax, in the same manner as bees do honey," if we understand by *worms*, insects not arrived at maturity; for the larva of Bombyx Mori, is also termed a silk worm, though it belongs to the moth tribe when perfect.

HEMIPTERA.

Croton Sebiferum—Poplar-leaved Croton, or Tallow-tree.—The Tallow-tree is not the natural food of the Wax insect, but as they mutually illustrate the same inquiry, they are represented in the same plate; and it is further presumed, that a short account of this useful plant will be deemed a proper sequel to the history of the insect.

Du Halde, when describing the Tallow-tree, says, " Il est de la hauteur d'une grande cerisier. Le fruit est renfermée dans un écorce qu'on appelle *Yen Kiou*, et qui s'ouvre par le milieu quand il est mûr, comme celle de la châtaigne. Il consiste en des grains blancs de la grosseur d'un noisette, dont la chair a les qualitez du suif; aussi en fait-on des chandelles, après l'avoir fait fondre, en y mêlant souvent un peu d'huile ordinaire, et trempant les chandelles dans la cire qui vient sur l'arbre dont je vais parler: il s'en forme autour du suif une espèce de croûte qui l'empêche de couler. Page 18. Vol. I.

Sir G. Staunton speaks nearly to the same effect: " From the fruit of the *Croton sebiferum*, of Linnæus, the Chinese obtain a kind of vegetable fat, with which they make a great proportion of their candles. This fruit, in its external appearance, bears some resemblance to the berries of the ivy. As soon as it is ripe, the capsule opens and divides into two, or, more frequently, three divisions, and falling off discovers as many kernels, each attached by a separate foot-stalk, and covered with a fleshy substance of a snowy whiteness, contrasting beautifully with the leaves of the tree, which, at this season, are of a tint between a purple and a scarlet. The fat, or fleshy substance, is separated from the kernels by crushing and boiling them in water. The candles made of this fat are firmer than those of tallow, as well as free from all offensive odour. They are not, however, equal to those of wax or spermaceti." This author further adds, " The wax for candles is generally the produce of insects, feeding chiefly on the privet, as is mentioned in the chapter of Cochin China. It is naturally white, and so pure as to produce no smoke; but is collected in such small quantities, as to be scarce and dear. Cheap candles are also made of tallow, and even of grease of too little consistence to be used, without the contrivance of being coated with the firmer substance of the tallow-tree or of wax." *Vide Chapter on Sou-choo-foo.*

The tallow-tree is now cultivated in the West Indies, where it thrives well, and produces fruit, and by proper attention may hereafter become useful.

NATURAL HISTORY OF THE INSECTS OF CHINA

Belostoma Indica?

HEMIPTERA.

BELOSTOMA INDICA?

Plate 18.

Sub-Order. Hemipterita, *Kirby.*
Section. Hydrocorisa, *Latreille.*
Family. Nepidæ, *Leach.*
Genus. Belostoma, *Latreille.* Nepa p. *Linn.*
Ch. Sp. B. " squalidè lutea, maculis fuscis, femoribus anticis nigro-lineatis, coxis quatuor posticis immaculatis." Long. Corp. 3 unc.
B. dirty clay coloured with brown spots, the anterior femora with black lines, and the four posterior coxæ immaculate. Length 3 inches.
Syn. Belostoma indica? *Enc. Méth.* X. *p.* 272.
Nepa grandis, *Donovan,* 1st edit. Stoll Cimic. 2. *t.* 7. *f.* 4. (Exclus. synon. Linn. Fabr. Merian, Rösel, and De Geer.)

M. Merian has given a plate and description of the South American Belostoma grandis in her work on the Insects of Surinam. We learn from that account, that in the larva and pupa state it lives in the water; that it is a voracious creature, and feeds not only on the weaker kinds of aquatic insects, but on some animals much larger than itself. The pupa* is represented on the back of a large frog in the water, and is designed to portray the manner in which it fastens on those creatures, holding them between its strong curved fore feet, and extracting the juices of their bodies through its singularly constructed beak. M. Merian says, the winged insect was produced from one of these pupæ on the twelfth of May, 1710.

Every writer on this insect since M. Merian appears indebted to her for their account of these few particulars; for though all the European species of the same family undergo precisely the same changes in their aquatic dwellings, among decayed vegetables, &c. at the bottom of the water, and quit it only in the winged state,† we are indebted to her for the time of the appearance of this exotic species in that state, as well as for a correct figure of its pupa.

Linnæus, following Merian, gives *Surinam* as the country of B. grandis; Margravius,

---

* The pupa is *semi-completa*: unlike the pupa of the Lepidoptera, &c. it scarcely differs in appearance or manners of life from the complete insect, but has only the rudiments of the wings. See the lower figure in the accompanying plate.

† Nepa cinerea and linearis are English species of this family; these live in the water till they have wings, when they occasionally quit it to pursue other winged creatures.

46   HEMIPTERA.

Brasil; and Fabricius, *America* generally. Donovan observed a slight difference between the Chinese specimen and the figures in preceding works referred to by Fabricius; but he nevertheless gave it as the Nepa Grandis of Fabricius, on the authority of the collection of the Right Hon. Sir Joseph Banks, Bart. The Asiatic species are, however, now regarded as specifically distinct from those of America, and hence I have given this doubtingly as the B. indica of Saint Fargeau.

BELOSTOMA (SPHÆRODEMA) RUSTICA.

Plate 19. fig. 1.

Ch. Sp.   B. rotundata, ecaudata, fusca, thoracis elytrorumque margine antico albido.   Long. Corp. lin. 7½.

B. round, without a tail, brown, with the margin of the elytra and the front of the thorax pale.   Length of the body 7½ lines.

Syn.   Nepa rustica, *Fabr. Ent. Syst.* 4. 62. 3.   (Exclus. syn. *Enc. Méth.* X. *p.* 273. et *Laporte Revis. Hemipt. p.* 18. Diplonychus rusticus.)   *Laporte op. cit. p.* 83.
Nepa plana, *Sulz. Hist. Ins. t.* 10. *f.* 2.   *Stoll Cim.* 2. *t.* 7. *f.* 6.

Insects in general discover an extraordinary degree of care and ingenuity in depositing their eggs in the most secure situations, or places where the infant brood, when hatched, may be provided with proper sustenance. Those of the aquatic kind usually lay them in recesses in the mud or sand, or under loose stones that lie at the bottom of the water: others, with as much care, and more ingenuity, hollow out the interior substance of the large stalks of water plants, and deposit their eggs in them; or, rising out of the water, lay them in the extreme branches of those plants, to secure them from other aquatic depredators. Belostoma rustica displays even more sagacity, or attachment for its eggs, than those creatures; for it never leaves them. Till they are hatched, it bears them on its back, in a cluster of an oval shape; these eggs are of an oblong form, and are fastened by the narrowest end to a thin film, or plate of cement, that causes them to adhere to the polished surface of the wing cases; when these eggs, about a hundred in number, are hatched, it casts off the exuviæ of the cluster, and differs no longer in general appearance from the male of the same species.

Our figures represent the situation of the eggs on the back, and the insect also after they are cast off. It is not commonly received with the eggs upon it. Found on the coast of Coromandel, as well as China.

1. *Belostoma rustica.*   2. *Nepa rubra.*

1. Callidea Dispar. 2. Stanbuschii.

HEMIPTERA.

NEPA RUBRA.
Plate 19. fig. 2.

GENUS. NEPA, *Linnæus.*
CH. SP. N. oblongo-ovata, depressa, fusca, longè bicaudata, abdomine supra rubro, linea nigrà. Long. Corp. (cauda exclus.) $1\frac{1}{3}$ unc.
N. oblong-ovate, depressed, brown, with two long filaments at the anus, abdomen above red, with a black dorsal line. Length $1\frac{1}{3}$ inch.
SYN. Nepa rubra, *Linn. Syst. Nat.* 2. 713. 2.   *Fabr. Syst. Rhyng. p.* 107.

CALLIDEA OCELLATA.
Plate 20. fig. 1.

SECTION. GEOCORISA, *Latreille.*
FAMILY. SCUTELLERIDÆ, *Leach.*
GENUS. CALLIDEA, *Laporte.* Chrysocoris, *Hahn.* Cimex p. *Linn.* Tetyra p. *Fabricius.*
CH. SP. C. carnea thorace scutelloque maculis flavescentibus, quibusdam puncto ocellari atro. Long. Corp. $\frac{3}{4}$ unc.
C. orange-red, thorax and scutellum with yellow spots, some of which have a black central spot. Length $\frac{3}{4}$ inch.
SYN. Cimex ocellatus, *Thunberg Nov. Sp. f.* 72.
Tetyra dispar, *Fab. Ent. Syst.* 4. *p.* 81.   *Syst. Rhyng. p.* 129.   *Stoll Cimic. t.* 37. *f.* 260.   *Donov. 1st edit.*

This curious insect is among the number of those lately brought from China. A figure of the upper surface is represented on a leaf of the *Camellia Sesanqua*, one of the vignette plates of Sir G. Staunton's History of the late Embassy to that country; a coloured figure and a short account of it may therefore prove acceptable to the readers of his volumes.

Stoll has also given a figure of it, and has also represented another sort, which he considers as the female *(letter A)*; it has no black points in the yellow spots of the thorax and scutellum: he mentions the Isle of Formosa as the native country of his specimens.

Fabricius observes, that it varies according to the sex by having an acute spine on each side of the thorax, as represented in the accompanying figures; this statement is, however, denied in the Encyclopédie Méthodique, vol. x. p. 410, by Messrs. Saint Fargeau

48   HEMIPTERA.

and Serville. The absence of these spines, although not perhaps a sexual character, certainly indicates a remarkable variety. I have certainly seen specimens destitute of these spines.

The pupa figured at the bottom of the plate on the left hand side is probably that of Tetyra Druræi or some allied species. Donovan is entirely silent respecting it.

ASTEMMA SCHLANBUSCHII.
Plate 20. fig. 2.

Family.	Lygæidæ.
Genus.	Astemma, *Laporte*. Lygæus, *Fabr*.
Ch. Sp.	A. sanguinea, thorace fasciâ abbreviatâ, scutello, elytrorum puncto alisque atris. Long. Corp. fere ¾ unc.
	A. red, thorax with an anterior abbreviated black band; scutellum, wings, and a spot on each wing-cover, black. Length nearly ¾ inch.
Syn.	Lygæus Schlanbuschii, *Fabr. Ent. Syst.* 4. p. 155. *Syst. Rh.* p. 222.

The accompanying figure does not represent the anterior transverse thoracic black band mentioned in the Fabrician description; the species varies in this respect. The pupa figured on the right hand side at the bottom of the plate is most probably that of this insect, although Donovan is entirely silent concerning it.

CALLIDEA STOCKERUS.
Plate 21. fig. 1.

Family.	Scutelleridæ.
Genus.	Callidea, *Laporte*. Cimex p. *Linn*. Tetyra p. *Fabr*.
Ch. Sp.	C. cærulea, thorace punctis 6, sc. 3 parvis anticis, 3 majoribus posticis; scutello 7, apiceque nigris. Long. Corp. lin. 6.
	C. blue, the thorax with 6 spots, three small in front, and three behind larger; the scutellum with 7 large spots, and the apex black. Length 6 lines.
Syn.	Cimex Stockerus, *Linn. Syst. Nat.* p. 715. *Fabr. Ent. Syst.* 4. p. 79.
	Cimex Stollii, *Wolff*.

This insect seems to be very common in China; for we rarely receive a parcel of the insects of China that does not include many of them. There are several distinct, but very

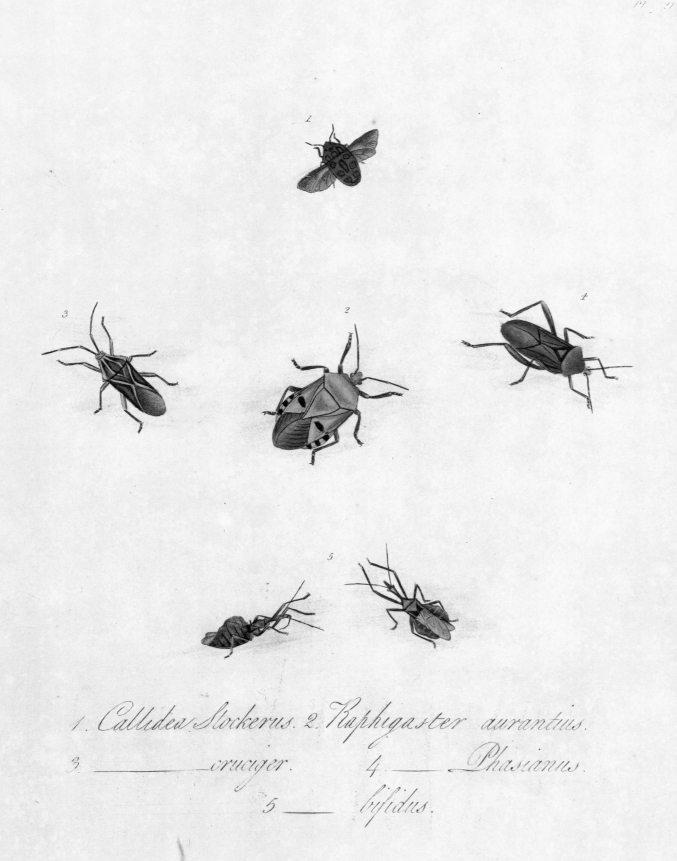

1. Callidea Stockerus. 2. Raphigaster aurantius.
3. ———— cruciger.   4. ——— Phasianus.
   5. ——— bifidus.

HEMIPTERA.

closely allied, species, which have been regarded as varieties, but which appear to be constant in their variations, being inhabitants likewise of distinct countries. These vary not only in their tints and the number of their spots, but also in their outline, which seems clearly to prove them specifically distinct. Donovan mentions one of these as " a charming miniature variety of C. Stockerus from *Africa æquin.* about one-third of the size of the Chinese specimens, and of a very deep blue colour. The marks both on the upper and under side precisely resemble those in the annexed figures."

RAPHIGASTER AURANTIUS.

Plate 21. fig. 2.

FAMILY.   PENTATOMIDÆ.
GENUS.   RAPHIGASTER, *Laporte.* Edessa p. *Fabr.*
CH. SP.   R. aurantius, capite thoracis margine antico, abdominis maculis marginalibus pedi-busque atris. Long. Corp. 1¼ unc.
          R. orange-coloured, with the head, anterior margin of the thorax, lateral spots of the abdomen, antennæ and legs, black. Length 1¼ inch.
SYN.   Cimex aurantius, *Fabr. Ent. Syst.* 4. p. 105. *Syst. Rhyng.* p. 149. (Edessa a.)
          Sulzer Ins. t. 10. *f.* 10. *Stoll Cimic, t.* 6. *f.* 39.

CHARIESTERUS CRUCIGER.

Plate 21. fig. 3.

FAMILY.   COREIDÆ, *Leach.*
GENUS.   CHARIESTERUS, *Laporte, Burm.* Lygæus p. *Fabr.*
CH. SP.   Ch. thorace acutè spinoso; oblongus, supra niger; thorace lineis, elytris cruce, ferrugineis. Long. Corp. lin. 10.
          Ch. with the angles of the thorax acutely spined; oblong, black above, thorax with two lines and the margins orange, elytra with an orange cross. Length 10 lines.
SYN.   Cimex cruciger, *Fabr. Mant. Ins.* 2. 289. 104.
          Lygæus cruciger, *Fabr. Ent. Syst.* 4. p. 141.

Donovan states, that this insect was " from the collection of Mr. Francillon, who received it from China. Fabricius describes it as a native of Brazil." This is certainly the true locality of the species. Donovan's figure was taken from a specimen which wanted the terminal joint of the antennæ.

HEMIPTERA.

CERBUS TENEBROSUS?

Plate 21. fig. 4.

GENUS.   CERBUS, *Hahn*.   Anisoscelis, *Latr*.   Lygæus p. *Fabr*.

CH. SP.   C. cinnamomeo-fuscus, antennis apice tibiisque dilutioribus.   Long. Corp. unc. 1.
C. cinnamon-brown, with the tips of the antennæ and tibiæ paler.   Length 1 inch.

SYN.   Lygæus tenebrosus? *Fabr. Ent. Syst. 4. p. 135. Burm. Handb. der Ent. 2. p. 340.*
Lygæus Phasianus, *Donov. 1st edit.*

Donovan gave this as identical with the Lygæus Phasianus, Fabr. from tropical Africa, although his specimens were brought from China by the late Mr. Ellis. This figure appears rather to represent a female of the common Lyg. tenebrosus.

HARPACTOR BIFIDUS.

Plate 21. fig. 5.

FAMILY.   REDUVIIDÆ.

GENUS.   HARPACTOR, *Laporte*.   Reduvius p. *Fabr*.

CH. SP.   H. ater, elytris fasciâ rufâ, scutello spinâ erectâ apice bifidâ.   Long. Corp. $\frac{9}{10}$ unc.
H. black, elytra with a red bar, scutellum with an erect spine forked at the extremity. Length $\frac{9}{10}$ of an inch.

SYN.   Riduvius bifidus, *Fabricius Ent. Syst. 4. 204. Syst. Rhyng. p. 285.*
Cimex bifidus, *Donovan, 1st edit.*

NATURAL HISTORY OF THE INSECTS OF CHINA

Papilio Paris.

Order. LEPIDOPTERA. *Linnæus.*

PAPILIO PARIS.

Plate 22.

Tribe.	Diurna, *Latreille.* (Papilio, *Linnæus.*)
Family.	Papilionidæ, *Leach.*
Genus.	Papilio, *Linnæus,* (Section, Equites.) *Latreille, Boisduval, &c.*
Ch. Sp.	P. alis nigris, aureo-viridi pulverulentis, posticis caudatis, maculâ (in utroque sexu) magna azureo-cæruleâ, ocello fulvo ad angulum ani, his subtus maculis septem marginalibus ocellatis. Expans. alar. 4 unc.
	P. with the wings black, powdered with golden-green atoms, the posterior with a broad tail and a large shining blue spot in both sexes, and a reddish eye at the anal angle; beneath with seven marginal eye-like spots. Expanse of the wings about 4 inches.
Syn.	Papilio Paris, *Linn. Syst. Nat.* 2. *p.* 745. *No.* 3. *Fabr. Ent. Syst.* 3. 1. *p.* 1. *No.* 1. *Drury Exotic Ins.* V. I. *t.* 12. *f.* 1. 2. *Cramer Pap.* 2. *pl.* 103. *f.* A. B. *Esper. Ausl. Schmett. t.* 2. *f.* 1. *Encycl. Méth.* IX. *p.* 69. *Boisduval Hist. Nat. Lep.* 1. *p.* 208.

The simile proposed by Linnæus, both for the arrangement and specific nomenclature of butterflies, is gleaned from ancient and fabulous history. The species are divided into sections of Trojan and Greek princes, heroes, deities, nymphs, and plebeians: and the species have received names in accordance with this fanciful theory, which, at least, in the writings of Linnæus is well conducted, and seems liable to less objection than the characters assigned to each section: for many species placed among the Equites, and a more considerable number with the Plebeii, are inconsistent with the essential criterion Linnæus has given. This arrangement has necessarily undergone material alterations in the *Entomologia Systematica* of Fabricius and other still more recent works; alterations certainly justified by the more comprehensive views now taken of this pleasing branch of Entomology. The Equites of Fabricius, with many additions, and a few exceptions, are the same as those in the two Linnæan sections: Papilio Priamus is, however, removed from the head of the *Equites Trojani,* and the precedence given to Papilio Paris.

Papilio Paris is an insect of considerable beauty. The general colour on the upper surface is obscure brown, nearly approaching black, but finely contrasted with brilliant green atoms, profusely sprinkled over it. The posterior wings are adorned with a large blue spot, which derives additional lustre from the dusky colour surrounding it. Another species, very similar to Papilio Paris, but without this spot, is also found in China. It has been supposed to be the female of our species, which opinion is adopted in the Encyclopédie Méthodique. Fabricius names it Bianor, after *Cramer, pl.* 103. *fig.* 6. Dr. Horsfield has described and figured another species, from Java, in his Lepidoptera Javanica, under the name of Papilio Arjuna, but which is so closely allied to Paris, that it may eventually prove to be only a geographical variety. Its larva is cylindrical, with a coriaceous shield-like plate, extending over the three anterior segments of the body; the chrysalis is greatly angulated, with the head notched. Another species, or at least strong variety, has been lately received from the Himalayan mountains.

PAPILIO CRINO.

Plate 23.

Ch. Sp.   P. alis nigris atomis viridi-aureis, fasciâ communi cæruleo-viridi; posticis caudatis, ocello anali rufo, his subtus lunulis viridibus cæruleis cinereisque. Expans. alar. $3\frac{1}{2}$ unc.

P. with the wings black and sprinkled with golden-green atoms, with a greenish-blue bar running across all the wings, the posterior pair tailed, with a red eyelet at the anal angle; beneath with green, blue, and ashy lunules. Expanse of the wings $3\frac{1}{2}$ inches.

Syn.   Papilio Crino, *Jones. Fabricius Ent. Syst.* 3. 1. *p.* 5. *Enc. Méth.* IX. *p.* 66. *Boisduval Hist. Nat. Lepid.* 1. *p.* 207.
Papilio Regulus, *Stoll Suppl. Cramer.* 5. *pl.* 41. *f.* 1.

This splendid butterfly is extremely rare, and its precise country is doubtful. Fabricius says, "Habitat in *Africa*. Mus. Dom. Drury." Donovan, however, who had access to Drury's collections, says, "We have found an unique specimen of this species in the collection of Mr. Drury, and on that authority we include it as a native of China. Fabricius erroneously gives Africa as its locality." In the Encyclopédie Méthodique, Africa is given. Boisduval gives "Indes orientales;" his unique specimen having been sent to him by M. Drege as from Cochin China, but which Boisduval thinks may possibly be erroneous. The manuscripts of Drury, now in my possession, throw no light upon the subject further than that there are several *unnamed* species indicated as inhabitants of China as well as of Sierra Leone. But from the strong affinity between Crino, Palinurus, Paris, &c. it is scarcely to be doubted that China or India is the real locality of Crino.

Renealmia exaltata, a majestic plant, near seven feet in height, bearing a fine pendant group of flowers at the summit, is figured in the plate.

Papilio Crino.

1. Papilio Coon. 2. Papilio Agenor.

PAPILIO COON.

Plate 24. fig. 1.

Ch. Sp. P. alis angustatis, anticis elongato-ovatis fuscis; posticis caudâ spatuliformi, atris, maculis baseos palmatis, lunulis submarginalibus albis maculâque duplici ad angulum ani flavâ. Expans. alarum 4½—5¼ unc.

P. with narrow wings, the anterior elongate ovate, brown on both sides; the posterior, with a spatulate tail, black, with palmated basal spots and submarginal lunules of a white colour; and two yellow spots at the anal angle. Expans. of the wings 4½—5¼ inches.

Syn. Papilio Coon, *Jones. Fabricius Ent. Syst.* 3. 1. *p.* 10. *Enc. Méth.* IX. *p.* 65. *Boisduval Hist. Nat. Lep. p.* 201.

Papilio Hypenor, *Enc. Méth.* IX. *p.* 65.

The original Fabrician description was derived from a specimen in the collection of Mr. Drury, and Donovan's figure is copied from the drawings of Mr. Jones, referred to by Fabricius. The translation of the Fabrician description of the lower wings is incorrectly rendered in the Encyclopédie Méthodique, and in consequence another description is given of a Javanese specimen of this species, under the name of P. Hypenor. It has recently been received in considerable numbers from Java, from whence I possess a specimen with the wings much longer and narrower than they are here represented.

PAPILIO AGENOR.

Plate 24. fig. 2.

Ch. Sp. P. alis nigris, basi sanguineis, anticis striatis, posticis dentatis, disco albo maculisque marginalibus atris. Expans. alar. 6 unc.

P. with the wings black, bloody at the base, the anterior with longitudinal paler markings, the posterior dentate with a white disc and black marginal spots. Expansion of the wings 6 inches.

Syn. Papilio Agenor, *Linn. Syst. Nat.* 2. *p.* 747. *Fabr. Ent. Syst.* 3. 1. *p.* 13. *Enc. Méth.* IX. *p.* 28. *Clerk Ic. t.* 15. *Cramer, pl.* 32. A.B. *Herbst. Pap. t.* 8. *f.* 3.

Papilio Memnon, ♀ *Boisduval Hist. Nat. Lepid. p.* 193.

This is one of the largest Chinese Papiliones we are acquainted with. The upper and under surfaces so nearly agree, that Donovan considered a figure of the first unnecessary. M. Boisduval has advanced several forcible reasons for regarding this and

LEPIDOPTERA.

several other allied insects (P. Laomedon, of Cramer, Anceus, Achates) as females of the very variable Asiatic species, Papilio Memnon; the caterpillar of which, according to Dr. Horsfield, is green, with the anterior segments narrowed and retractile, the third being elevated at the top and marked with an eye-like spot on each side. It feeds upon the species of Citrus.

The plant figured is *Plumbago Rosea*. Rose-coloured Lead-wort.

PAPILIO PERANTHUS.

Plate 25.

Ch. Sp.   P. alis nigris, supra basi cærulescenti-viridibus, subtus apice pallidis, posticis obtuse dentatis, caudatis, subtus lunulis rufescentibus serie digestis. Expans. alar. 4 unc.

P. with the wings black, greenish blue at the base; beneath paler at the external margins, posterior pair dentate and tailed, with red lunules on the under side arranged in a transverse series. Expansion 4 inches.

Syn.   Papilio Peranthus, *Fabricius Ent. Syst.* 3. 1. *p.* 15. *Enc. Méth.* IX. *p.* 66. *Boisduval Hist. Nat. Lepid.* 1. *p.* 203.

The original Fabrician description was made from a specimen from Cochin China in the Banksian collection. Donovan mentions another which came from Canton, and M. Boisduval gives Borneo, Java, and Celebes, as its localities.

The insect is represented on a small twig of *Arundo Bambos (Bamboo* or *Cane)*, a well known plant, mentioned by Sir G. Staunton as being one of the most useful productions of China.

PAPILIO TELAMON.

Plate 26. fig. 1.

Ch. Sp.   P. alis caudatis, concoloribus, flavescentibus, maculis fasciisque nigris, posticis utrinque strigâ sanguineâ nigromarginatâ. Expans. alar. 3 unc.

P. with the wings coloured alike pale yellowish, with black spots and bands, the posterior with very long narrow tails, and a red streak bordered with black at the anal angle. Expansion of the wings 3 inches.

Syn.   Papilio Telamon, *Donovan. Boisduval Hist. Nat. Lep.* 1. *p.* 250.

The singular delicacy and beauty of this Papilio is not the only claim it has to the particular attention of Entomologists: it is clearly an undescribed species; and perhaps

Papilio Peranthus.

1. Papilio Telamon.   2. Papilio Agamemnon.

LEPIDOPTERA.

the only specimen of it yet brought to Europe is that from which our figure is copied. It was taken near Pekin, by a gentleman in the suite of Earl Macartney, in the embassy to China; and was originally in the possession of Mr. Francillon, of London, who kindly permitted drawings and descriptions to be made of this and every other insect in his magnificent collection that could enhance the value of this publication. It is still so rare that M. Boisduval states that he had never seen a specimen of it.

Papilio Telamon bears a distant resemblance to P. Protesilaus, but a much stronger to P. Ajax: pursuing then the metaphorical method of arranging the butterflies in the Linnæan manner, the name of the father of Ajax, who was one of the distinguished Grecian Princes at the siege of Troy, has been given to this species.

PAPILIO AGAMEMNON.

Plate 26. fig. 2.

Ch. Sp. P. alis nigris viridi-maculatis, posticis breviter caudatis, his subtus ocello lunato maculisque rubris. Expans. alar. 3½ unc.

P. with the wings black and spotted with pale green, the posterior pair with short tails and ornamented beneath with a lunate eyelet and red spots. Expansion of the wings 3½ inches.

Syn. Papilio Agamemnon, *Linn. Syst. Nat.* 2. p. 748. *Fabr. Ent. Syst.* 3. 1. p. 33. *Enc. Méth.* IX. p. 46. *Boisduval Lep.* 1. p. 230.
Papilio Ægistus, *Cramer*, 106. C. D. (*corrected, p.* 151.)

Papilio Agamemnon is found in several parts of Asia (China, Bengal, Java, the Moluccas and Philippine Islands, Manilla, Timor). The under side is beautifully adorned with a number of bright green spots of various sizes. The general colour is pale pink, diversified with shades of chestnut brown. The upper side is much plainer; the general colour is black, except the spots, which are green, and precisely agree in shape with those on the under side. Dr. Horsfield has figured the transformations of this insect (Lepid. Javan. pl. 4. f. 12), the larva is short and thick, with a forked tail; the chrysalis has the head very obtuse. This species is the type of Dr. Horsfield's second section of the genus, having the club of the antenna oval and compressed.

PAPILIO PROTENOR. ♀

Plate 27.

Ch. Sp. P. alis anticis fuscis nigro-striatis; posticis dentatis, nigris, atomis pallidis, maculá duplici rufâ anguli ani.   Expans. alar. 5½ unc.

P. with the anterior wings brown with black longitudinal stripes, the posterior dentate, black with pale atoms, and a double red spot at the anal angle.   Expans. of the wings 5½ inches.

Syn. Papilio Protenor, *Fabricius Ent. Syst.* 3. 1. 13. *Cramer, pl.* 49. A. B. *Enc. Méthod.* IX. *p.* 30.   *Boisduval Hist. Nat. Lepid. p.* 198.
♀ Papilio Laomedon, *Jones. Fabricius Ent. Syst.* 3. 1. 12.   *Donovan, 1st edition,* (*nec Cramer, pl.* 50. *f.* A. B.)

The collection of the late Mr. Latham contained the original specimen from which Mr. Jones' drawing (referred to by Fabricius) was made. The present figure was copied, by Mr. Jones' permission, from that drawing.

PAPILIO EPIUS.

Plate 28. fig. 1.

Ch. Sp. P. alis nigris, flavo maculatis; posticis dentatis fasciâ irregulari, maculis adjectis flavis, maculâque anguli analis rufâ.   Expans. alar. fere 3½ unc.

P. with the wings black, spotted with yellow; the posterior pair dentate with an irregular yellow bar, accompanied externally with additional spots and a large red spot at the anal angle.   Expanse of the wings nearly 3½ inches.

Syn. Papilio Epius, *Jones. Fabricius Ent. Syst.* 3. 1. *p.* 35. *Enc. Méth.* IX. *p.* 43. *Boisduval Hist. Nat. Lepid. p.* 238.
Papilio Erithonius, *Cramer, pl.* 232. A. B.
Papilio Demoleus, *Esper. Ausl. Schmett, tab.* 50. *f.* 1—4.

Papilio Epius and Papilio Demoleus are so similar in their marks and colours, that many authors have confounded one species with the other. Papilio Epius is chiefly distinguished by the red spot in the interior margin of the lower wings, having no blue eye-shaped mark above it.

Papilio Protenor.

1. Papilio Epius. 2. Papilio Demoleus.

Morpho Rhetenor.

LEPIDOPTERA.

PAPILIO DEMOLEUS.

Plate 28. fig. 2.

Ch. Sp.   P. alis nigris, flavo-maculatis, posticis dentatis fasciâ flavâ subrectâ ocelloque anali
dimidiatim cæruleo rufoque.   Expans. alar. 3½ unc.
P. with black wings spotted with yellow, the posterior dentate with a nearly straight
and regular yellow fascia, and an ocellus at the anal angle blue above and red be-
neath.   Expanse of the wings 3½ inches.
Syn.   Papilio Demoleus, *Linn. Syst. Nat.* 2. 753.   *Fabr. Ent. Syst.* 3. 1. 34.   *Kleeman
(Rosel, add.) t.* 1. *f.* 2. 3.   *Cramer Ins. t.* 231. *f.* A. B.   *Boisduval Hist. Nat.
Lep.* 1. *p.* 237.   *Encycl. Méth.* IX. *p.* 43.
Papilio Demodocus, *Esper. Ausl. Schmett. t.* 51. *f.* 1.

Linnæus gives the Cape of Good Hope as the habitat of this species; and Boisduval also mentions the coast of Guinea, Senegal, and Madagascar. Fabricius, however, particularly says, " Habitat in Indiæ orientalis Citro, Dr. Koenig," describing the larva as solitary, smooth, of a yellowish green colour, with a reddish head, two tentacles on the neck, and a bifid tail. Boisduval has, however, applied this observation to P. Epius, stating that P. Demoleus had been reared at Senegal by M. Dumolin, and that its larva feeds on the citron trees.

MORPHO RHETENOR.

Plate 29.

Family.   Nymphalidæ, *Swainson.*
Genus.   Morpho, *Fabricius (Syst. Gloss. in Illig. Mag.)*
Ch. Sp.   M. alis suprà nitidissimè cyaneis; subtùs umbrino griseoque variis, ocellis cæcis.
Expans. alar. 5½ unc.
M. with the wings on the upper side dazzling cyaneous blue, beneath varied with
umber and grey, with blind eyelets.   Expanse of the wings about 5½ inches.
Syn.   Papilio Rhetenor, *Cramer, pl.* 15. A. B.   *Herbst. Pap. t.* 27. *f.* 1. 2.   *Esper. Pap.
Exot. t.* 42. *f.* 1.   *Sulzer Ins. t.* 13. *f.* 1.   *Enc. Méth.* IX. *p.* 444.

Whatever effect the artist can produce by a combination of the most brilliant colours employed in painting, must be far surpassed by comparison with the dazzling appearance of this splendid creature. It is impossible to find in any part of the animal creation colours more beautiful or changeable. Pale blue is the principal colour, but new tints meet the eye in every direction, varying from a silvery green to the deepest purple; and the whole surface glittering with the resplendence of highly polished metal.

This splendid species was long confused with Papilio Menelaus of Linnæus, the authors of the Encyclopédie Méthodique, however, cleared up the confusion, proving them to be quite distinct. The species, and indeed the entire group to which it belongs, are, however, natives of South America. Sulzer, indeed, states that his specimen came from China, and evidently on this authority Donovan introduced the species into this work.

Thea Laxa (Bohea, or broad-leaved Tea), is figured in the plate. Sir G. Staunton says, the bohea tea is supplied in China from the province of Fochen: the green tea from Kiang-nan. The leaves of these teas vary in some degree in form according to the age of the plant; those of the bohea are the broadest; Thea stricta has much longer leaves, they are lanceolated, and more deeply serrated than those of the bohea. Many authors have considered them varieties of the same species. It flowers in England in August and September.

ACRÆA VESTA.

Plate 30. fig. 1.

FAMILY.   HELICONIIDÆ, *Swainson.*
GENUS.   ACRÆA, *Fabricius.* (Heliconia p. *Fabricius olim.*)
CH. SP.   A. alis oblongis integerrimis, utrinque corticinis; omnium supra limbo posteriori fusco serieque punctorum interrupto. Expans. alar. 2½ unc.
A. with the wings oblong and entire, of a pale yellow brown; with a dark brown border in which are white spots. Expanse of the wings 2½ inches.
SYN.   Papilio (Helic.) Vesta, *Jones. Fabr. Ent. Syst.* 3. 1. *p.* 163. *Enc. Méth.* IX. p. 233.
Papilio Terpsichore, *Cramer Pap. pl.* 298. *f.* A. B. C.

Papilio Vesta is the only insect of the *Heliconii* division of Butterflies described by Fabricius as peculiar to China, in his *Ent. Syst.*, the majority being inhabitants of Africa. It is a rare species. The *Papilio Vesta* of Cramer is a very different insect, being the *P. Erato* of Fabricius.

1. Acræa Vesta. 2. Pieris Pasithoe.
3. Pieris Hyparete.

PIERIS PASITHOE.

Plate 30. fig. 2.

FAMILY. PAPILIONIDÆ, *Leach.*
GENUS. PIERIS, *Fabricius, Boisduval* (Papilio, Heliconia, *Linn.*)
CH. SP. P. alis suboblongis, nigris, supra cærulescenti-albo-maculatis; posticis subtus disco flavo, nigro venoso, fasciàque basali ferruguinea.   Expans. alar. 3½ unc.
P. with oblong wings of a black colour spotted on the upper side with bluish white, the posterior pair with the disc beneath yellow, with black veins and a broad red basal fascia.   Expanse of the wings 3½ inches.
SYN. Papilio (Helicon.) Pasithoe, *Linn. Syst. Nat.* 2. 755.   *Fab. Ent. Syst.* 3. 1. 179.
Pieris P. *Encycl. Méth.* IX. *p.* 148.   *Boisduval Hist. Nat. Lep.* 1. *p.* 451.   *Drury Illust. Exot. Ent.* 2*nd edit. v.* 2. *p.* 16.
Papilio (Dan. Cand.) Dione, *Drury.* 1*st edit.*
Papilio Porsenna, *Cramer, pl.* 43. *fig.* D. E. *and pl.* 352. *fig.* A. B.

PIERIS HYPARETE.

Plate 30. fig. 3.

CH. SP. P. alis suboblongis, integerrimis, albis, utrinque apice, subtus venis nigris, posticis subtus, plus minusve flavis, maculis sanguineis limbo nigro apicali digestis. Expans. alar. 3 unc.
P. with the wings rather oblong, entire, white with a black margin on both sides and with black veins beneath; the posterior on the under side more or less stained with yellow, with a row of six red spots in the black border.   Expanse of the wings about 3 inches.
SYN. Papilio (Helicon.) Hyparete, *Linn. Syst. Nat.* 2. *p.* 763. (nec *Fabr. Ent. Syst.* 3. 1. *p.* 176.   *Enc. Méth.* IX. *p.* 153.   *Boisduval Hist. Nat. Lep.* 1. *p.* 455.
Papilio Autonoe, *Cramer,* 187. C. D. et 320. A. B.

Several distinct, but nearly allied, species have been confounded together under the name of Hyparete; Donovan observed, " We have two sorts of this species; one with the marginal row of red spots on the posterior wings disposed in a deep border of black; the other has the red spots on a whitish ground.   They are certainly the two sexes of Papilio Hyparete.   Found near Canton, in China."   The sexes do not, however, vary in this respect, the latter individuals mentioned in this passage are therefore most probably the

LEPIDOPTERA.

P. Eucharis of Drury (Epicharis, Enc. Méth. and Boisduval; Hyparete, Fabricius), or the P. Autonoe of Cramer and Boisduval.

The leaf represented in the plate is that of *Sophora Japonica (Shining-leaved Sophora)*, an elegant and valuable timber tree, of which Sir G. Staunton speaks as very frequent in China.

PIERIS (IPHIAS) GLAUCIPPE.

Plate 31. fig. 1.

Sub-Gen. Iphias, *Boisduval.*

Ch. Sp.   P. alis supra albis, anticis maculâ magnâ apicali (medio fulvo) nigrâ, subtus (nisi dimidio basali anticarum) cinereis, strigis minutis fuscis irroratis. Expans. alar. 4 unc.

P. with the upper sides of the wings white, the anterior having a large black spot at tip, the middle of which is rich orange; beneath, except the basal half of the anterior wings, greyish with brown waves. Expanse of the wings 4 inches.

Syn.   Papilio (Dan. Cand.) Glaucippe, *Linn. Syst. Nat.* 2. 762. *Fabr. Ent. Syst.* 3. 1. p. 198. *No.* 618. *Herbst. Pap. Fab.* 96. *f.* 1—3. *Enc. Méth.* IX. *p.* 119. *Drury Exot. Lepid.* 1. *pl.* 10. *f.* 3. 4. *2nd edit. p.* 20. *Horsfield Lep. Jav. p.* 130. (Colias Gl.)

♀ Pap. Callirhoe, *Fab. Mant. Ins.* 2. 20. 215.

Iphias Glaucippe, *Boisduval Hist. Nat. Lep.* 1. *p.* 596.

P. Glaucippe is an elegant insect: very common in China, and it is said, in some adjacent parts of Asia, also Bengal, Java, &c. The Papilio Callirhoe of Linnæus is considered as the female of this species: few authors deem it more than a variety (β). Dr. Horsfield has described and figured the caterpillar and chrysalis of this species in his work upon the Lepidoptera of Java above referred.

1. Pieris Glaucippe. 2. Pieris Pyrene.

1. Colias Pyranthe.   2. Colias Philea.

LEPIDOPTERA.

PIERIS (THESTIAS) PYRENE?

Plate 31. fig. 2

SUB-GEN. THESTIAS, *Boisduval.*

CH. SP.   P. " alis integerrimis, rotundatis, flavis; primoribus apice (medio fulvo) nigris; subtus nebuloso-maculatis." *Linn. loc cit. infra.* Expans. alar. fere 2½ unc.

P. with the wings entire, rounded, yellow; the anterior with a large black spot, the centre of which is bright orange; beneath with cloud-like marks. Expanse of the wings about 2½ inches.

SYN.   Papilio (Dan. Cand.) Pyrene, *Linn. Syst. Nat.* 2. 762. 86? *Enc. Méth.* IX. *p.* 120. Boisduval *Hist. Nat. Lep.* 1. 593. (Thestias P.) *Drury Exot. Ent.* 2nd edit. 1. *p.* 11. *pl.* 5. *f.* 2.

Papilio Sesia, *Fabricius Ent. Syst.* 3. 1. *p.* 203. *Donovan,* 1st. edit.

The insect here figured, judging at least from the upper side, agrees with the Linnæan description of P. Pyrene, the habitat of which is given by Linnæus as China. Fabricius, however, who refers to Linnæus, gives America as its locality. It is now satisfactorily ascertained that it inhabits China and various parts of the East Indies.

The plant figured is *Limodorum Tankervillæ,* an elegant and much admired production of China.

COLIAS (CALLIDRYAS) PYRANTHE.

Plate 32. fig. 1.

GENUS.   COLIAS, *Fabricius.* Papilio (Danai Candidi), *Linnæus.*

SUB-GEN.   CALLIDRYAS, *Boisduval.*

CH. SP.   C. alis integerrimis, rotundatis, albis, puncto apiceque nigris; subtus cinereo undulatis puncto fulvo. Expans. alar. fere 3 unc.

C. with the wings entire, rounded, and white, having a discoidal spot and the margin black; beneath with ashy waves and a fulvous spot. Expanse of the wings nearly 3 inches.

SYN.   Papilio Pyranthe, *Linn. Syst. Nat.* 2. *p.* 763. *Enc. Méth.* IX. *p.* 97. *Boisduval Hist. Nat. Lep.* 1. *p.* 611.

Papilio Gnoma of Fabricius; P. Alcyone, Cramer; P. Nepthe, Fabricius; and P. Chryseis, Drury (1st edition); are in all probability varieties of this insect.

COLIAS (CALLIDRYAS) PHILEA.

Plate 32. fig. 2.

Ch. Sp. C. alis integerrimis, subangulatis, flavis; anticis macula, posticis limbo, luteis. Expans. alar. 3¾ unc.

C. with the wings entire, somewhat angulated, bright yellow, with a large discoidal spot in the anterior and a broad margin of the posterior pair orange. Expanse of the wings 3¾ inches.

Syn. Papilio (Dan. Cand.) Philea, *Linn. Syst. Nat.* 2. 764. *Fabricius Ent. Syst.* 3. 1. *p.* 212. *Cramer Pap. pl.* 173. E. F. *Roesel Ins. Bel. t.* 3. *f.* 5. *Boisduval Hist. Nat. Lep.* 1. *p.* 619. (Callidryas P.)

Linnæus says of this species, "Habitat in Indiis," which may be either taken for the East or West Indies; Roesel calls it " die *indianische* goldborte;" and Donovan states that his specimen was received from China. The real locality, however, not only of this butterfly, but of all the species of the section to which it belongs, is South America and the West Indies. M. Boisduval gives the Argante of Hubner, Lolia of Godart, Aricia of Cramer, Melanippe Cramer, and Larra of Fabricius, as varieties of the female of this species.

The plant represented in the plate is *Melastoma Chinensis*.

MORPHO (DRUSILLA) JAIRUS.

Plate 33.

Family. Nymphalidæ, *Swainson*.
Genus. Morpho, *Fabricius*. (Papilio, *Dan. Festiv. Fabr. olim.*)
Sub-Gen. Drusilla, *Swainson Zool. Illust.* 1. *pl.* 11. (Hyades, *Boisduval Hist. Nat. Lep.* 1. *pl.* 13. *f.* 1.)
Ch. Sp. M. alis integris, fuscis; posticis disco baseos albo, supra oculo maximo, subtus duobus dissitis. Expans. alar. 4½ unc.

M. with the wings entire, brown; the posterior, with the basal disc white, with a single large eye on the upper and two on the lower side. Expanse of the wings 4½ inches.

Syn. Papilio F. Jairus, *Fabr. Ent. Syst.* 3. 1. *p.* 54. *Cramer Pap. pl.* 6. A. B. and *pl.* 185. A. B. C. *Enc. Méth.* IX. *p.* 445.
Papilio Cassiæ, *Clerk Icon. tab.* 29. *fig.* 3.

A specimen of this extremely rare Butterfly was contained in the collection of Dr. Hunter, now the property of the University of Glasgow; a fragment in the British Mu-

Morpho Jairus.

Nymphalis Bernardus

LEPIDOPTERA.

seum; and one in fine preservation in the collection of Mr. Francillon. Except these, and the specimens from which the figures in the annexed plate are copied, Donovan had never seen it in any cabinet whatever. It had been figured only by two authors, Clerk in his *Icones insectorum rariorum*, and Cramer in his *Papillons exotiques*. The figures of Clerk and Cramer do not strictly agree: we observe those of the first much lighter coloured, and the white space on the upper wings considerably larger than in any of the figures in Cramer's plates.

Fabricius says it is a native of the East Indies. One specimen figured by Cramer was brought from the isle of Amboyna. It seems therefore not peculiar, like some insects, to China.

NYMPHALIS (CHARAXES) BERNARDUS.

Plate 34.

GENUS. NYMPHALIS, *Latreille.* (Papilio Nymphalis, *Fabricius.*)

SUB-GEN. CHARAXES, *Boisduval.* (Jasia, *Swainson.*)

CH. SP. N. alis fulvis, anticis apice atris, fasciâ mediâ flavâ, posticis caudatis strigâ punctorum ocellatorum. Expans. alar. 3½ unc.

N. with fulvous wings, the anterior black at the tips with a broad pale yellow band, the posterior tailed, with a row of black ocellated spots. Expanse of the wings 3½ inches.

SYN. Papilio (Nymph.) Bernardus, *Jones. Fabricius Ent. Syst.* 3. 1. *p.* 71.

This uncommonly rare Chinese butterfly has not been figured in any other work. Fabricius described it only from the drawings of Mr. Jones. I possess a specimen in which the central fascia is nearly white, and is continued half way across the posterior wings, and the black spots in the latter are very broad and confluent, without white in the centre.

The plant represented is the *Camellia Japonica (Japan Rose)*, a native of Japan and China, which blossoms from January to May. It is a lofty and magnificent plant, rising to the height of several feet: there is a variety of it with double flowers, perfectly white; and another in which the flowers are variegated with white and red.

LEPIDOPTERA.

ARGYNNIS ERYMANTHIS.

Plate 35. fig. 1.

GENUS. ARGYNNIS, *Fabricius.*

CH. SP. A. alis subrotundatis, subdentatis, fulvis, anticis fasciâ flavescenti transversâ mediâ nigro-maculatâ, apice nigris; posticis serie punctorum duabusque lunularum nigrarum. Exp. alar. 2—3 unc.

A. with the wings rather rounded and indented, fulvous, the anterior with a transverse pale yellow fascia, spotted with black, the tips black; the posterior wings with a row of black spots and two rows of narrow spots. Expanse of the wings from 2 to 3 inches.

SYN. Papilio (Dan. Fest.) Erymanthis, *Drury Exot. Ent. vol.* 1. *pl.* 15. *f.* 3. 4. *Cramer, pl.* 238. *f.* 9. *Fabr. Ent. Syst.* 3. 1. *p.* 139. *Enc. Méth.* IX. *p.* 257.
Papilio Lampetia, *Cramer Pap. pl.* 148. *fig.* E.

It is the rarity, and not the beauty of this butterfly, which induced Donovan to add it to this selection. It is probably far from common in China, being very seldom sent to Europe among the insects of that country.

CYNTHIA ORITHYA.

Plate 35. fig. 2.

GENUS. CYNTHIA, *Fabricius.* (Papilio Nymphales gemmati, *Linn.*)

CH. SP. C. alis denticulatis, supra nigris aut fuscis, singularum ocellis duobus iride fulvâ; anticis costâ strigisque apicalibus albis, his subfalcatis; posticis rotundatis. Expans. alar. 2. unc.

C. with the wings dentate, the anterior subfalcate, the posterior rounded; above blacker brown shaded with blue, each having two eyelets with a fulvous circle; the anterior margin and several apical fasciæ white. Expanse of the wings 2 inches.

SYN. Papilio N. Orithya, *Linn. Syst. Nat.* 2. *p.* 770. (nec *Abbot* and *Smith Lep. Georg.* v. 1. t. 8.) *Roesel Ins.* 4. *t.* 6. *f.* 2. *Cramer Pap. pl.* 19. C. D. 32. E. F. 281. E. F. 290. A. B. C. D.
Papilio N. Orythia, *Fabr. Ent. Syst.* 3. 1. *p.* 91. *Donovan,* 1st *edit.*

Donovan observes, that "the varieties of Papilio Orythia are numerous, and seem to differ according to climate of the countries of which they are natives. It is common in North America, Jamaica, India, &c. The variety from North America is almost wholly

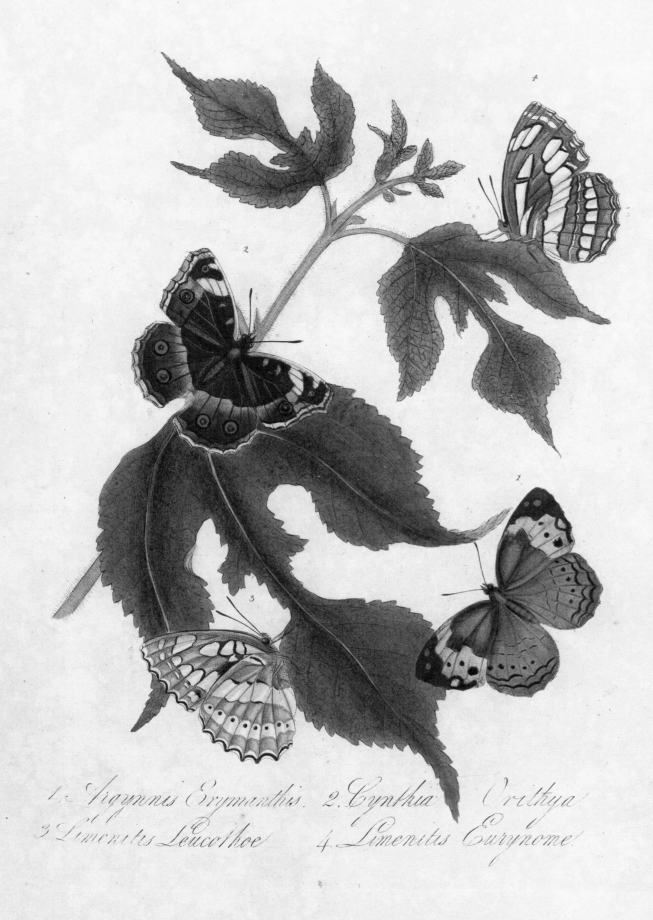

1. Argynnis Erymanthis. 2. Cynthia Orithya.
3. Limenitis Leucothoe. 4. Limenitis Eurynome.

LEPIDOPTERA.

brown, and those from Jamaica have less blue in the disk of the lower wings than those from China." Donovan, however, clearly here mistook distinct species for varieties. The American species thus named by Abbot and Smith is the P. Larinia of Fabricius, and in my copy of the Entomologia Systematica, which belonged to Professor Weber, the companion of Fabricius, the words " India orientali " are introduced in lieu of Jamaica.

Papilio Clelia of Cramer, which is found on the coast of Guinea, has been supposed a variety of Papilio Orythia. Fabricius, in the *Entomologia Systematica*, has made it a distinct species. It greatly resembles P. Orythia, but has no more blue colour on the posterior wings than is concentrated in a large spot near the base.

LIMENITIS LEUCOTHOE.

Plate 35. fig. 3.

GENUS.   LIMENITIS, *Fabricius*. (Papilio Nymphales Phalerati, *Linn.*)

CH. SP.   L. alis dentatis, supra fusco-nigris, subtus fulvis; utrinque fasciis tribus macularibus albis; posticarum fasciâ intermediâ punctis nigris antrorsum fœtis. Expans. alar. $2\frac{1}{2}$ unc.

L. with the wings dentate, above brownish black, beneath clay coloured; with three rows of white spots on both sides, the intermediate fascia of the posterior wings with black spots towards the base. Expanse of the wings $2\frac{1}{2}$ inches.

SYN.   Papilio Leucothoe, *Linn. Syst. Nat.* 2. *p.* 780. *Fabr. Ent. Syst.* 3. 1. *p.* 129. *Enc. Méth.* IX. *p.* 430. *Herbst. Pap. tab.* 240. *f.* 5. 6.

Papilio Polyxina, *Donov. 1st edition.*

Donovan regarded this as a new species, giving the following as the true Leucothoe of Linnæus. The description given by that author, and especially his notice of the third row of spots in the posterior pair of wings being composed " ex maculis 7 albis puncto nigro fœtis," clearly applies to this and not to the following insect.

K

LEPIDOPTERA.

LIMENITIS EURYNOME.

Plate 35. fig. 4.

CH. SP.   L. alis dentatis supra fusco-nigris, subtus fulvis; fasciis interruptis macularibus albis, subtus fusco cinctis, anticis fasciâ longitudinali baseos è maculis duabus triangularibus compositâ.   Expans. alar. 2½ unc.

L. with the wings dentate, above brownish black, beneath fulvous, with interrupted white maculated bands, which on the under side are edged with brown, the anterior have also a longitudinal basal fascia, composed of two triangular white spots, the bases of which are opposed to each other.   Expanse of the wings 2½ inches.

SYN.   Limenitis Eurynome, *Westw.*
Papilio Leucothoe, *Donov. 1st edit.*
Papilio Aceris major ex India, *Esper. Pap. tab.* 82. *f.* 1.

CYNTHIA ŒNONE.

Plate 36. fig. 1.

CH. SP.   C. alis denticulatis supra luteis margine omni nigro; posticis basi late nigris, maculâ cyaneâ.   Expans. alar. 2½ unc.

C. with the wings denticulated, above pale clay-coloured, with all the margins black, the base of the posterior black with a large cyaneous blue spot.   Expanse of the wings 2½ inches.

SYN.   Papilio (Nymph. Gemm.) Œnone, *Linn. Syst. Nat.* 2. 770. " alis denticulatis, primoribus albido maculatis subbiocellatis, posticis basi cyaneis ocellis duobus." *Fabr. Ent. Syst.* 3. 1. 90. *Kleman Ins.* 1. *t.* 3. *f.* 1. 2.
Vanessa Œnone, *Enc. Méth.* IX. p. 318.

Donovan says, that this insect is found throughout Asia (which is the locality assigned to it by Linnæus and Fabricius), and is very common in China. In the Encyclopédie Méthodique, the Cape of Good Hope is given as its habitat. The Linnæan specific character is applicable to the female; the male (according to M. Godart), which is here figured, having no eyes on the upper side of the wings.

1. Cynthia Oenone. 2. Cynthia Almana.
3. Nymphalis Lubentina.

LEPIDOPTERA.

CYNTHIA ALMANA.

Plate 36. fig. 2.

Ch. Sp. C. alis anticis falcatis, posticis intus subcaudatis, omnibus supra fulvis, ocellis sesqui-altero, subtus fuscescentibus, posticis lineâ flavidâ transversâ mediâ. Expans. alar. 2¾ unc.

C. with the anterior wings falcate, the posterior subcaudate at the inner angle, all fulvous above, with an ocellus on each; beneath brownish, the posterior with a yellowish transverse line in the centre. Expanse of the wings 2¾ inches.

Syn. Papilio (N. G.) Almana, *Linn. Syst. Nat.* 2. 769. *Fabr. Ent. Syst.* 3. 1. *p.* 89. *Cramer Pap. pl.* 58. F. G. *Herbst. Pap. t.* 172. 1. 2.

The angulated form of the wings of this butterfly gives it a remarkable appearance. The eyes on the wings somewhat resemble those of the Peacock butterfly, to which, in some other respects, it bears no distant similitude. It is common in China; Fabricius gives its *habitat* Asia.

NYMPHALIS (ACONTHEA) LUBENTINA.

Plate 36. fig. 3.

Ch. Sp. N. alis subdentatis, fusco-virescentibus; anticis utrinque fasciâ albâ, maculari; posticis apice punctis chermisinis, serie duplici digestis. Expans. alar. 2½ unc.

N. with the wings subdentate, brownish-green, the anterior on each side with a row of white spots, the posterior with scarlet spots arranged in a double series towards the extremity. Expanse of the wings 2½ inches.

Syn. Papilio (Nymph.) Lubentina, *Fabricius Ent. Syst.* 3. 1. *p.* 121. *Enc. Méth.* IX. 400. *Cramer Pap. pl.* 155. C. D. *Herbst. Pap. t.* 146. 1. 2.

Aconthea Lubentina, *Horsfield Lep. Jav. pl.* 5. *f.* 5.

Papilio Lubentina is figured only in the works of Cramer: his specimen is not precisely like ours, but agrees in all the essential peculiarities, and is unquestionably the same species. The semitransparent spots on the anterior wings are much larger in Cramer's figure than in the insect before us.

68 LEPIDOPTERA.

NYMPHALIS JACINTHA.

Plate 37. fig. 1.

Ch. Sp. N. alis repando-dentatis, fuscis; anticis striga punctorum alborum, posticis apice albis margine fusco lunulis albis. Expans. alar. 4—4½ unc.

N. with the wings scalloped, brown; the anterior with a row of spots on the anterior pair at the tips, posterior externally white, the margin being brown with white lunules. Expanse of the wings from 4 to 4½ inches.

Syn. Papilio (Nymph. Phal.) Jacintha, *Drury, app. vol.* 2. *pl.* 21. *f.* 1. 2. *Fabricius Ent. Syst.* 3. 1. *p.* 60. ♀
Papilio (Nymph.) Liria, *Fab. Ent. Syst.* 3. 1. *p.* 126. ♂?
Papilio Perimale, *Cramer, pl.* 65. C. D. 67. B.

This curious butterfly was found in the province of Pe-tche-lee, in China. It is in all probability the female of Pap. (N.) Liria, Fabr.

It is represented, with P. Antiochus, on a leaf of the *Urtica Nivea (White Nettle).**

NYMPHALIS ANTIOCHUS.

Plate 37. fig. 2.

Ch. Sp. N. alis supra holosericeo-nigris, fasciâ communi nitide aurantiâ; anticarum abbreviatâ. Expans. alar. 2¾ unc.

N. with the wings above holosericeous black, with a broad shining orange bar common to all the wings, but abbreviated in the anterior pair. Expanse of the wings 2¾ inches.

Syn. Papilio (Dan. Fest.) Antiochus, *Linn. Mant.* 1. 537. *Drury, app. vol.* 3. *pl.* 7. *f.* 3. 4. *Fabricius Ent. Syst.* 3. 1. *p.* 44. *Enc. Méth.* IX. *p.* 409.
Papilio Eupalemon, *Cramer, tab.* 143. *f.* B. C. Le Velouté. *Daubenton, pl. enl.* 68. *f.* 3. 4.

This insect is very rare in European cabinets of insects. The specimen figured by Drury came from the Brazils, and Cramer's from Surinam. Fabricius, however, describes

---

* Sir G. Staunton speaks of a cloth that the Chinese manufacture from the fibres of a dead nettle. Query, Is this the species employed for that purpose? no other is noticed by that author in the lists of plants collected in China. The nettle is of general use in Russian Tartary also; the *Kuriles*, and other Siberian tribes, make cloth, cordage, thread, &c. of it. *Gordon*, &c.

1. Nymphalis Jacintha. 2. Nymphalis Antiochus.

Nymphalis Sylla

LEPIDOPTERA.

it as a native of China; and Donovan states that the insect figured in the collection of drawings of Mr. Jones, of Chelsea, was a native of China, as well as the specimen in his own collection. There must, however, have been some mistake in respect to these specimens, for not only are all the immediately allied species natives of South America, but Stoll observed its transformations in that country, and says, that the caterpillar feeds on the tamarind; it is green, with two long spines on the head, and numerous other shorter spines on the body.

NYMPHALIS SYLLA.

Plate 38.

Ch. Sp. N. alis dentatis, supra nigris viridi-maculatis striatisque; anticis fasciâ maculari niveâ. Expans. alar. 3¾ unc.

N. with the wings dentate, black on the upper side, with green spots and lines, the anterior with a row of white spots, the central fascia of the posterior wings externally radiated. Expanse of the wings 3¾ inches.

Syn. Papilio Sylla, *Cramer Pap. t.* 43. *f.* F. G.
Papilio Sylvia, *Herbst. Pap. t.* 247. *f.* 2. 3.
Papilio (N.) Gambrisius, *Fabr. Ent. Syst.* 3. 1. *p.* 85. *Donov. 1st edit.*
Nymphalis Sylvina, *Enc. Méth.* IX. *p.* 381.

A specimen of this very rare Papilio was taken in one of the small islands on the eastern coast of China, and was in the possession of Mr. Francillon. Sir J. Banks, Bart. had a specimen of it from another part of the East Indies. It also occurs in Java and Amboyna (Enc. Méth.).

MYRINA (LOXURA) ATYMNUS.

Plate 39. fig. 1.

FAMILY. LYCÆNIDÆ, *Swainson.*
GENUS. MYRINA, *Fabr. Enc. Méth.*
SUB-GEN. LOXURA, *Horsfield Lep. Jav. p.* 119.
CH. SP. M. alis supra fulvo-testaceis, apice nigro, posticis longè caudatis. Expans. alar. 1½—1¾ unc.
M. with the wings above orange red, with the tips black, the posterior with very long tails white at the tips. Expanse of the wings 1½—1¾ inches.
SYN. Papilio (Pleb. Rural.) Atymnus, *Fab. Ent. Syst.* 3. 1. *p.* 283. *Cramer, pl.* 331. *fig.* D. E. (palpis deteritis.)
Loxura Atymnus, *Horsfield Lep. Jav. p.* 119. *pl.* 2. *f.* 6. *Boisduval Hist. Nat. Lep.* 1. *pl.* 7. *fig.* 3.

This is also a scarce species. Donovan's specimen was from the collection of the late Duchess Dowager of Portland, who procured it from China. Another specimen in the cabinet of Sir J. Banks, Bart. is from Siam. Dr. Horsfield found it in Java.

The plant represented in the plate is *Hemerocallis Japonica*, brought from China by Mr. Slater.

THECLA MÆCENAS.

Plate 39. fig. 2.

GENUS. THECLA, *Fabricius.* (Papilio Hesperia Pleb. rural., *Fabricius olim.*)
CH. SP. Th. alis bicaudatis atris disco cæruleo, subtus brunneo nebulosis. Expans. alar. 1¾ unc.
Th. with the wings black and furnished with two tails, the disc being blue; beneath clouded with brown. Expanse of the wings 1¾ inches.
SYN. Hesperia (R.) Mæcenas, *Jones. Fab. Ent. Syst.* 3. 1. *p.* 271. 45. *Enc. Méth.* IX. *p.* 639.

1. *Myrina Atymnus.*   2. *Thecla Mœcenas.*

1. Deilephila Nechus. 2. Glaucopis Polymena.

LEPIDOPTERA.

DEILEPHILA NECHUS?

Plate 40. fig. 1.

SECTION.   CREPUSCULARIA, *Latreille.*
FAMILY.   SPHINGIDÆ, *Leach.*
GENUS.   DEILEPHILA, *Ochsenh.* Sphinx p. *Linn. &c.*
CH. SP.   D. " alis integris; anticis viridibus; strigâ testaceâ; posticis nigris; maculis baseos fasciaque flavis." *Fabric. loc. cit. subtus.* Expans. alar. 3½ unc.
       D. with the wings entire, the anterior green with a testaceous streak, (" with testaceous marks," *Donov.*) the posterior black with spots at the base, and a row of spots near the extremity. Expanse of the wings 3½ inches.
SYN.   Sphinx Nechus? *Fabricius Ent. Syst.* 3. 1. *p.* 377. *Cramer Ins. t.* 178. *f.* B.

The number of Chinese species of this genus, already described, is very limited: the insect represented in the accompanying figures is the largest of them; but as this is inferior in size to several kinds found in Europe, we conceive there must remain many larger species of the genus unknown to collectors of foreign insects, and yet very common in China. In the latter part of Sir G. Staunton's work, that author mentions the larva of a *Sphinx Moth* which furnish an article for the table of the Chinese. We regret that the indefinite expression cannot assist us to determine the species, and scarcely the genus, of the insect alluded to.*

The specimen figured in the annexed plate was in the collection of Mr. Francillon, who received it from China. The habitat of D. Nechus, given by Fabricius, is America; and Cramer has represented a small variety of the same species from North America.

---

* European naturalists are entirely ignorant of the Chinese insects in the state of larva and pupa, if we except a few species of the Cimices, Cicada, and some altogether uninteresting insects, that have been accidentally brought among others from that country. Hence it must remain undetermined whether they correspond in form with those of other parts of the world. It is, however, highly probable, from their great affinity to those in the perfect state, that in the state of larva they may also agree. The extensive collection of the larvæ of sphinges made by Mr. Abbot in North America affords no singularly constructed animal distinct from those found in Europe; they vary indeed in their colours, but preserve uniformly the characters found on the same genus in other countries. We noticed among the drawings of the late Mr. Bradshaw the figure of a Chinese sphinx, apparently *S. Hylas*, together with a larva similar to that of the S. Stellatarum: it was green, and, like all the known larvæ of the family (except the *Adscitæ* division), was perfectly free from hairs: it was also furnished with a horn at the posterior part of the body.

72   LEPIDOPTERA.

Moreover, Donovan's figure and description do not precisely correspond with the Fabrician description, so that on both these grounds I have considered it advisable to give the specific name with a mark of doubt. Sphinx Batus and Sphinx Gnoma are nearly allied to this insect, particularly the former; both are found in different parts of the East Indies.

GLAUCOPIS POLYMENA.

Plate 40. fig. 2.

FAMILY.	ZYGÆNIDÆ, *Leach*.
GENUS.	GLAUCOPIS, *Fabr.* Sphinx, *Linn. Donov.*
CH. SP.	G. nigra alis maculis luteis, anticis tribus, posticis duabus; abdomine cingulis duobus coccineis. Expans. alar. fere 2 unc.
	G. black, wings spotted with deep yellow, the anterior having three and the posterior two spots, abdomen with two scarlet bands. Expanse of the wings nearly 2 inches.
SYN.	Sphinx Polymena, *Linn. Syst. Nat.* 2. 806. *no.* 40. *Cram. Ins. t.* 13. *f.* D. *Fabr. Ent. Syst.* 3. 1. *p.* 396. *Drury Exot. Ent.* 1. *t.* 26. *f.* 1.

This beautiful creature is probably scarce in China; at least it is very rarely found among the insects brought from that country.

It is figured on the plate with the *Rosa semperflorens (Ever-blowing China Rose).*

SESIA HYLAS.

Plate 41. fig. 1.

GENUS.	SESIA, *Fabricius.* Sphinx, *Linn. Donov.*
CH. SP.	S. alis fenestratis, abdomine barbato viridi, cingulo purpureo. Expans. alar. 2½ unc.
	S. with transparent wings, body pale yellow green, abdomen with a brush at the tip and a purple belt round the middle. Expanse of the wings 2½ inches.
SYN.	Sphinx Hylas, *Linn. Mant.* 1. 539. *Fabr. Ent. Syst.* 3. 1. *p.* 379.
	Sphinx Picus, *Cramer Ins. t.* 148. *f.* B.

1. Sesia Hylas. 2. Callimorpha? Thallo.
3. Callimorpha? ruficollis. 4. Callimorpha? bifasciata.

LEPIDOPTERA.

CALLIMORPHA? THALLO.

Plate 41. fig. 2.

SECTION. NOCTURNA?
FAMILY. ARCTIIDÆ, *Stephens?*
GENUS. CALLIMORPHA? *Latreille.* Sphinx, *Linn.*
CH. SP. C. alis oblongis integerrimis nigris anticis fasciis duabus, posticis unica flavis; capite rubro. Expans. alar. 2 unc.
C. with oblong entire black wings, anterior pair shaded with blue at the base, and with two pale yellow fasciæ; posterior wings with a pale yellowish space, head red. Expanse of the wings 2 inches.
SYN. Papilio Thallo, *Linn. Syst. Nat.* 2. 756. *Fabr. Ent. Syst.* 3. 1. p. 173.
Sphinx pectinicornis, *Linn. Syst. Nat.* 2. 807. *Fab. Ent. Syst.* 3. 1. p. 399. *Edwards Aves*, 36. t. 226.
Phalæna tiberina, *Cramer*, t. 32. *f.* C. D.
Sphinx Thallo, *Donov. 1st edit.*

Donovan entered into a lengthened observation, shewing that Fabricius had given a Papilio Thallo in all his works when no such Papilio was in existence, and that Edwards' figure of the insect in question was derived from a mutilated or mended specimen. The former error is, however, rather to be attributed to Linnæus, who introduced all the confusion by describing Edwards' figure both as a Papilio and Sphinx.

The figures of Cramer and Edwards do not precisely agree; in the former, the disk of the posterior wings is yellowish, with a deep border of black: in the other, the yellow occupies only a space near the base, and forms a semi-lunar mark near the anterior margin of those wings. Donovan suspected, with Cramer, that they are but the two sexes of one species. Cramer says both his specimens came from China, from whence Donovan's insects were also received. The real affinities of this and some other allied species are very perplexing, they seem, however, to connect the Zygænidæ with the Arctiidæ.

L

LEPIDOPTERA.

CALLIMORPHA? RUFICOLLIS.

Plate 41. fig. 3.

Ch. Sp. C. " alis integerrimis nigro-purpurascentibus, fasciâ communi maculisque duabus flavis, thorace antice brunneo." Expans. alar. 1¼ unc.
C. with the wings entire, black purple, a semicircular yellowish band communicating across all the wings, and two spots of the same colour near the apex, collar reddish. Expanse of the wings 1¼ inch.

Syn. Sphinx ruficollis, *Donov. 1st edit.*

This and the following species were considered by Donovan to be undoubted nondescripts: both species were in the collection of Mr. Francillon, who received them from China.

CALLIMORPHA? BIFASCIATA.

Plate 41. fig. 4.

Ch. Sp. C. alis fulvis; anticarum fasciâ apiceque nigris. Expans. alar. 1 unc.
C. with orange or fulvous wings, anterior pair with a black bar across the middle, and the tips black. Expanse of the wings 1 inch.

The plant represented is *Thuja Orientalis (China Abor-vitæ Tree)*, an ornamental evergreen, much esteemed by the Chinese, and very frequently represented in their landscapes. Sir G. Staunton remarks, in the account of the journey from Pekin to Canton, that great quantities of this plant grew to a prodigious height in the valley in which stands the city of Yen-choo-foo.

NATURAL HISTORY OF THE INSECTS OF CHINA

Saturnia Atlas.

SATURNIA ATLAS.

Plate 42.

FAMILY. BOMBYCIDÆ.
GENUS. SATURNIA, *Schranck.* (Phalæna Attacus, *Linn.*)
CH. SP. S. alis anticis falcatis, luteo variis, macula fenestrata anticis sesquialtera. Expans. alar. 8 unc.

S. with the anterior wings falcate; yellow brown, varied with paler markings, each wing with a triangular talc-like spot in the middle, the anterior having also a smaller one near the tips. Expanse of the wings 8 inches.

SYN. Phalæna Attacus Atlas, *Linn. Syst. Nat.* 2. *p.* 808. *Petiv. Gaz. t.* 8. *f.* 7. *Fab. Ent. Syst.* III. 1. *p.* 407. *Cram. Ins.* 381. C. 382. A.

The nocturnal Lepidoptera are remarkable for the neatness and simplicity of their colours. Their elegancies consist in the infinite variety and delicacy of intermingled tints: the contrast of spots, specklings, and lineations, which constitute the minutiæ of insect beauty. Some species are to be excepted in this remark; the larger kinds are often gaudy, and the smallest exhibit a display of the richest colours, fancifully disposed, and most elegantly diversified.

The European species are numerous, and pretty well ascertained: those of remote countries remained, at the period of the publication of the first edition of this work, and still remain in great obscurity. The species inhabiting China are almost unknown;* for Fabricius describes not more than twenty species in all the cabinets in Europe. From this scanty number a few are selected to illustrate the genus, and if these appear deficient in point of interest or variety, it may stimulate others to collect new species whenever an opportunity occurs. The moths, not only of China, but of every country except Europe, have received but little attention. In Europe, the number of this tribe exceeds that of any other: on the contrary, the extra European species are comparatively the most inconsiderable of our acquisitions. The Papiliones, or butterflies, are a showy and lively race: they sport in the open fields by day, and attract the traveller's curiosity; hence our cabinets abound with them. But the moths, infinitely more numerous, and not less pleasing, are seldom seen; in the gloominess of their dispositions, they seek the

---

* *Fab. Ent. Syst.* These are chiefly described from insects in the collection of "*Monson Londini*," of which no figures are extant, and the collection unknown.

LEPIDOPTERA.

obscurity of the forest in the day, and only venture on the wing when the sun is down. In Europe we visit their nocturnal haunts without difficulty or dread; but in hotter climates these are oftentimes impenetrable, or the lurking places of ferocious animals; and few will expose themselves to their attacks to increase the catalogue of exotic moths.*

Phalæna Atlas is the first species we have to notice. It is one of the largest of the moth tribe,† and is, indeed, a gigantic creature. The species is common, but not peculiar to China, being found in other parts of Asia, and in America. The influence of climate is easily traced in the individuals from different countries; that from Surinam is the largest, and of the deepest colours. The Chinese kind is the next in size; the colours incline to orange, and the anterior wings are more falcated, or hooked, at the ends. We

---

* The far greater number of moths can only be taken in the woods at night. This is termed *mothing* by collectors. The moths begin to stir about twilight, and when almost dark, commence their flight. The collector is furnished with a large gauze folding-net, in which the insects are caught indiscriminately, for it is impossible to distinguish one species from another, and often is so dark, that the object itself can barely be discerned. Different species have their favourite haunts, some the lanes and skirts of woods, but many of them prefer the open breaks in the most retired places. As it would be unsafe, or impossible, to penetrate the woods in many countries, it is better to collect the larvæ, or caterpillars, for these may be found on the trees in the day-time, and if kept in little gauze cages, and carefully fed, will change into chrysalides, and produce the moths. This is certainly tedious, and few travellers will divert their attention from more important observations; but were they to appropriate their leisure to this branch of science, they would materially improve entomology. Mr. Abbot investigated a small district of Georgia, in North America, in this manner, and our cabinets are indebted to his labours for several hundred species, altogether new in Europe. The reader may estimate the importance of these discoveries, by referring to the two expensive volumes of North American Lepidopterous Insects; and reflecting, that the originals of all the species included in that work are but a small selection from those he has furnished us with. Viewing these as the result of one man's research, in an inconsiderable portion of North America, what a variety of new and splendid kinds would be the reward of those, who should explore the more genial regions of Asia, Africa, and South America, with equal diligence and information!

We have hazarded an assertion which may seem inadmissible, that the Phalænæ are infinitely more numerous than the Papiliones, or any other tribe of insects. Not that we possess more, but because, in every country that has been investigated, experience justifies such opinion. For instance, in Great Britain we have only sixty[a] Papiliones, and by mere accident two or three local species have lately been added; of the Phalænæ we have more than 900. The same comparative proportion is observed throughout the countries of the European continent; and it is singularly analogous, that our opinion is confirmed, by the recent discoveries of Mr. Abbot in America.

† When Linnæus described it, few of the very large species of Phalæna were known. There are two species from the interior of Africa, which are larger than the Chinese Atlas, and several others scarcely inferior in magnitude.

---

[a] (There are now about eighty-five indigenous British butterflies, and between seventeen and eighteen hundred moths. J. O. W.)

LEPIDOPTERA.

have two other Asiatic *varieties* still smaller, with the wings extremely falcated. These are to be regarded as distinct species.

The larva of Phalæna Atlas is figured by *M. Merian*, in the *Insecta Surinamensia*, plate 52: it is about four inches in length, green, with a yellow stripe disposed longitudinally. Upon each segment are four distinct round tubercles, of a coral-like orange colour, which are surrounded with very delicate hairs. The pupa is large, and inclosed in a web of an ochre colour. The silk of this web is of a strong texture, and it has been imagined, that if woven, it would be superior in durability to that of the common silk worm. *Seba* has also represented the larva at fig. 1. plate 57. vol. 4. *Thesaurus Naturæ*. It is nearly six inches in length, and bulky in proportion; the Phalæna is also larger than that figured by Merian, which is a small specimen of the Surinam species. According to Merian, there are three broods of this insect in a year; they are very common, and feed on the orange trees. Linnæus says, they adhere so tenaciously to the leaves that they can scarcely be taken off.*

The common silk worm, or Phalæna Mori, belongs to this family, and merits observation as a native of China. The art of weaving its threads into silk is of the earliest date. The discovery is attributed to the *Seres*, a people of the East Indies, supposed the Chinese.† In the days of Solomon, we are told, a woman named Pamphilia, of the Island of Co, was skilled in the art of making cloth of the silk brought from the country of the Seres. The most ancient of the Chinese writers ascribe the invention to one of the women of the emperor *Hoang ti*, named *Si ling*, and in honour *Yuen fei*.‡ When Rome degenerated into voluptuousness, Persia, its dependency, furnished this article of luxury; but it is supposed they were indebted to the Chinese for it, and being supplied only in small quantities, it was consequently dear. In Rome it was so scarce, as to be worn only by persons of the first distinction.

The Chinese historians affirm, that the discovery was considered at first of such importance, that all the women in the palace of the emperor were engaged in rearing the insect and weaving its silk. In after times, the silk of China was a principal article of commerce; but latterly, its value has been materially lessened by the culture and fabrication of silk in other countries. As the Chinese know little of the use of linen, the silk is a staple article of their own consumption. The jesuit missionaries mention several

---

* Larva verticillata verrucis pilosis nec folliculos grandes, tenaces, vix extricandos. *Linn. Syst. Nat.*
† Velleraque ut foliis depectant tenuia Seres. *Virg. Georg.* II. 122.
‡ Du Halde, *Des Soyeries*. Les plus anciens écrivans de cet empire en attribuent la découverte à une des femmes de l'Empereur *Hoang ti*, nommée *Si ling*, et surnommée par honneur *Yuen fei*.

sorts of it in use among the Chinese; some admired for beauty, and others for durability. It is generally supposed these are not merely the effect of different manufacture, but are the produce of distinct insects.* Sir G. Staunton speaks of the culture of silk worms

---

* M. Merian says, in the description of the Surinam variety of Phalæna Atlas: "Telam ducunt fortem, quare bonum fore sericum rata, istius aliquam collegi copiam et in Belgium transmisi, ubi eadem optima judicata est: ut itaque, si quis Erucas istas congregandi laborem non detrectaverit, et bonæ notæ bombycem, et maximum hinc lucrum sibi comparare posset." The thread of which this coccon's web is composed is so strong, that it has been imagined it would make good silk. I have brought some of it into Holland, which has been esteemed such; so that if any one would take the trouble to collect a number of these caterpillars, they would be found good silk worms, and produce great profit. *Merian.*—Abbot informs us, the Moths of the Emperor tribe in general are called silk worms by the people of Georgia; and in the description of Phalæna Cecropia is still more explicit: for he says, "the caterpillar spins on a twig; the outside web is coarse, the inner covered with silk, like a silk worm's coccon. It is said this silk has been carded, spun, and made into stockings, and that it will wash like linen." *Abbot's Ins. by Dr. J. E. Smith.*—These insects are all of the same natural order, P. Cecropia is rather smaller, but very similar to P. Atlas, and this information at least corroborates the assertion of Merian.

An opinion that the Chinese rear several kinds of insects for the sake of their silk has long been prevalent. Dr. Lettsom proposes a query on this subject, "Which species of moth or butterfly is it, the caterpillar of which, in China, affords that strong grey kind of silk, and how is it manufactured or wore? How are these silk worms or caterpillars preserved, fed, and managed? The introduction of such a new silk into England would be a useful acquisition, and redeem entomology from the censure it is now branded with, of being a mere curiosity void of any real utility."[a] If *Lesser* and *Lyonet* are to be relied on, the *Théologie des Insectes* answers this query. "At this day there are to be found in China, in the province of Canton, silk worms in a wild state, which, without any care being taken of them, make in the woods a kind of silk which the inhabitants afterwards gather from the trees. It is grey, without lustre, and is used to make a very thick and strong cloth, named there Kien Tcheon. It may be washed like linen cloth, and does not stain." A gentleman resident in the East Indies speaks of a large Phalæna producing silk in that country: "We have a beautiful silk worm north-east of Bengal, that feeds on the Ricinus, whence I call it Phalæna Ricini; it is sea-green, with soft spines, very large and voracious, and spins a coarse, but strong and useful silk. The moth is of great size, with elegant dark plumage. Is it known to European naturalists?" *In a collection of papers published by Dr. Anderson in Madras*, 1788, 1789.—M. Le Bon, Reaumur, Roesel, and several others, have attempted to weave the silk of spiders as a substitute for that of silk worms, but their experiments rather amuse and point out the ingenuity of the proposers than promise to be useful; for after many trials, it appears that the silk of spiders would be inferior in lustre and far more expensive than that of silk worms. Sir G. Staunton alludes to these experiments in his description of the Java forests. "In some open spots were found webs of spiders, woven with threads of so strong a texture, as not easily to be divided without a cutting instrument; they seemed to render feasible the idea of him who, in the southern provinces of Europe, proposed a manufacture from spiders' threads, which is so ridiculous to the eyes of those who have only viewed the flimsy webs such insects spin in England." Many other substances of a soft texture have also been wrought into a variety of trifling articles, as gloves, stockings, &c. of the fibres of Asbestos earth, or mountain flax, beard of the large *Pinna* shell, &c. &c.

---

[a] Naturalist and Traveller's Companion, 1774.

1. Heleona milataris.   2. Eusemia lectrix.

LEPIDOPTERA.

in China, but only of the common sort. It will gratify curiosity, if not prove advantageous, should future observers ascertain what kind of insects the Chinese appropriate to making silk, and whether P. Atlas is of the number, as has been conjectured. It is indeed to be observed, that in India several distinct species of Saturnia are known to be employed in the production of silk; the most important of which are the Tusseh (S. Paphia, Linn.), the Arrindi (S. Cynthia, Drury), and the Kolisurra silk worm of the Dukhun.*

HELEONA MILITARIS.

Plate 43. fig. 1.

FAMILY. ARCTIIDÆ?
GENUS. HELEONA, *Swains. Zool. Illustr. N. Ser.* 116.
CH. SP. H. alis patulis concoloribus luteis apice maculisque violaceis, anticis extus albo-maculatis. Expans. alar. $3\frac{1}{2}$ unc.
H. with the wings extended at rest, the anterior and posterior pairs coloured alike, luteous yellow, with the extremity and spots at the base violet, the anterior with whitish spots at the tips. Expanse of the wings $3\frac{1}{2}$ inches.
SYN. Phalæna militaris, *Linn. Syst. Nat.* 2. 811. *Fabricius Ent. Syst.* 3. 2. p. 416. *Roesel. Ins.* 4. t. 6. f. 3. *Cramer. Ins.* t. 29. f. B.

The natural situation of this and some other allied insects is doubtful; it forms the type of Mr. Swainson's group Heleona, but is considered by that author to belong to the tribe of Sphingides, and family of Zygæidæ (Anthoceridæ, Swainson).

---

* See the Memoirs of Dr. Roxburgh in the Linnæan Transactions, and of Lieut.-Col. W. H. Sykes in the Transactions of the Royal Asiatic Society.

EUSEMIA LECTRIX.

Plate 43. fig. 2.

GENUS. EUSEMIA, *Dalman.* Phalæna, *Linn, &c.*

CH. SP. E. alis incumbentibus nigris, maculis cæruleis flavis albisque, posticis rubro alboque maculatis. Expans. alar. 3 unc.

E. with the wings incumbent, black, anterior with blue, yellow, and white spots, posterior with red and white spots. Expanse of the wings 3 inches.

SYN. Phalæna (Noctua) lectrix, *Linn. Syst. Nat.* 2. p. 834. *Fabr. Ent. Syst.* 3. 1. p. 475.

Eusemia lectrix, *Dalm. Monogr. Castn.*

This is so scarce an insect, that Mr. Drury informed Donovan he had only been able to procure a single specimen in the course of thirty years collecting insects.

EREBUS MACROPS.

Plate 44. fig. 1.

FAMILY. NOCTUIDÆ.

GENUS. EREBUS, *Latreille.* Thysania, *Dalman.*

CH. SP. E. alis dentatis, fuscis, nigro-undulatis; anticis ocello magno luteo, annulo nigro cincto. Expans. alar. 5¼ unc.

E. with the wings dentated, brown with black waves, the anterior having a large luteous ocellus, surrounded by a black ring. Expanse of the wings 5¼ inch.

SYN. Phalæna (Attacus) Macrops, *Linn Syst. Nat.* 4. p. 225.

Noctua Bubo, *Fabr. Mant. Ins.* 2. 209. *Ent. Syst.* III. 1. p. 9. *Donovan,* 1st *Edit.* (Phalæna B.) *Sulzer. Ins. t.* 22. *f.* 2. *Cramer. Pap. t.* 171. *f.* B.

This is the largest of the Chinese *Noctuæ;* some very similar species, but without the orange eye, and of a smaller size, are peculiar to China.

1. Erebus Macrops. 2. Hipparchus zonarius. 3. Callimorpha? Panthorea.

LEPIDOPTERA.

HIPPARCHUS ZONARIUS.

Plate 44. fig. 2.

FAMILY. GEOMETRIDÆ, *Leach*.
GENUS. HIPPARCHUS, *Leach*.
CH. SP. H. alis viridibus, margine posteriore late rufescente, singulis maculâ marginali viridi. Expans. alar. 1½ unc.
H. with the wings green, deeply bordered with pale red, with a green spot on the exterior margin of each wing. Expanse of the wings 1½ inch.
SYN. Phalæna Zonaria, *Donov. 1st edit.*

CALLIMORPHA? PANTHOREA.

Plate 44. fig. 3.

GENUS. CALLIMORPHA?
CH. SP. C. alis cæruleo-nigris, fasciâ maculari apicis albâ. Expans. alar. 2½ unc.
C. with blue-black wings, having a row of white spots along the posterior margins. Expanse of the wings 2½ inches.
SYN. Phalæna Panthorea, *Cram. Ins. t.* 322. *f.* C.
Phalæna pagaria, *Fab. Ent. Syst.* III. 2. *p.* 153. *Donov. 1st edit.* (Phalæna, Geometra, p.)

The insect here figured, and those represented in plate 41. fig. 2. and plate 43. fig. 1. are very intimately allied together; nevertheless, Donovan separated them, placing one in each of the three great divisions, Sphinx, Bombyx, and Geometra.

Order. NEUROPTERA. *Linnæus.*

ÆSHNA CLAVATA.

Plate 45. fig. 1.

FAMILY. LIBELLULIDÆ, *Leach.*
GENUS. ÆSHNA, *Fabr.* Libellula p. *Linn. Donov.* Cordulegaster, *Leach.*
CH. SP. Æ. abdomine clavato, basi gibbo; corpore nigro, fusco viridique variegato. Expans. alar. 3½ unc.
Æ. with the abdomen clavate gibbose at the base; body black, varied with brown and green; stigma brown. Expanse of the wings 3½ inches.
SYN. Æshna clavata, *Fabr. Ent. Syst.* II. *p.* 385. *Spec. Ins.* 1. *p.* 526. 4.

Linnæus divides the dragon flies (Libellula, Linn) into two sections:—" 1. alis patentibus acquiescentes;" and "2. (alis erectis) oculi distantes remotique." Fabricius divides the Linnæan Libellulæ into three distinct genera; the first retains the Linnæan name, the second and third are called Aeshna and Agrion. Their most essential characters are taken from the form and situation of the mouth, and therefore require a deep magnifier to determine them with accuracy. Donovan states, that he had examined those parts in the greater number of the species Fabricius has described, and found his characters agree, except in one instance; which Donovan nevertheless considered a proof of the impracticability of adopting the whole of his system: he describes *Libellula Chinensis*, and refers to the only figure that has been given of it, in one of the plates of Edwards's Natural History of Birds, 1745.* Had Fabricius ever seen and examined this rare species, he must have referred it to his genus *Agrion*, each of the lips being bifid, or two-cleft, as in Libellula virgo and puella,—the essential characteristic of the genus Agrion; for the mouths of the Libellulæ of Fabricius differ altogether in structure, and are not notched in the slightest degree, as Libellula clavata, ferruginea, 6-maculata, and the European species, Libellula depressa, will sufficiently illustrate.

Donovan, however, rejecting the Fabrician generic distribution, states, that Æshna clavata must be arranged with L. grandis and forcipata; but it is nearer allied to Cordulegaster annulatus, Leach (Libellula Boltoni of Donovan's British Insects.)

---

* That Fabricius should have erred in the location of a species which he had never seen, but knew only through a rude figure, is not surprising; but surely such a circumstance can be no proof of the impropriety of a system founded, as Donovan clearly shews, on characters of stability. J. O. W.

1. Aeshna clavata. — 2. Libellula variegata.
3. Libellula 6 maculata.

NEUROPTERA.

LIBELLULA VARIEGATA.

Plate 45. fig. 2.

GENUS. LIBELLULA, *Linn. &c.*
CH. SP. L. alis fusco-maculatis et undulatis, basi flavis, posticis versus apicem maculâ magnâ fuscâ puncto flavo; apice albo. Expans. alar. 3 unc.
L. with the wings spotted and undulated with brown, yellow at the base, the posterior towards the tips with a large brown fascia, having a small yellow spot, the tips yellow. Expanse of the wings 3 inches.
SYN. Libellula variegata, *Linn. Am. Acad.* 6. 412. 86. *Syst. Nat.* 1. 2. p. 904. *Drury Exot. Ent.* 2nd edit. 2. p. 94. nec *Fabr. op. cit.* p. 382.
Libellula Histrio, *Fabr. Mant. Ins.* 1. 337. 24.
Libellula Indica, *Fabr. Ent. Syst.* 2. 376. *Guérin Icon. R. An. Ins.* pl. 60. f. 1. Donovan, 1st edit.
Libellula Arria, *Drury Exot. Entom.* 1st edit. v. 1. pl. 46. f. 1.

Another species of Libellulidæ peculiar to India, and found in China, greatly resembles this insect; it is probably a variety of it.

LIBELLULA 6-MACULATA.

Plate 45. fig. 3. ♂ et ♀

CH. SP. L. alis anticis maculis tribus costalibus atris; ultima stigmate niveo; posticis fasciis flavescentibus. Expans. alar. 1½ unc.
L. with three black spots on the costa of the anterior wings, stigma white, posterior wings with yellow clouds. Expanse of the wings 1½ inch.
SYN. Libellula 6-maculata, *Fab. Ent. Syst.* 2. p. 381.

These delicate insects appear to be male and female; they are almost a miniature resemblance of the two sexes of Libellula depressa found in Europe; one having the abdomen yellow, and the other blue.

NEUROPTERA.

AGRION CHINENSIS.

Plate 46. fig. 1.

GENUS. AGRION, *Fabr.* Calepteryx, *Leach.*
CH. SP. A. alis anticis testaceo-obsoletis, posticis viridibus, apice fuscis. Expans. alar. $2\frac{3}{4}$ unc.
A. with the anterior wings brownish, posterior green with brown tips. Expanse of the wings $2\frac{3}{4}$ inches.
SYN. Libellula Chinensis, *Linn. Syst. Nat.* 2. 904. 15. *Fabr. Ent. Syst.* 2. 379. *Edw. Aves t.* 112. *Guérin Icon. R. An. Ins. t.* 60. *f.* 4. (Agrion c.)

The only two specimens of this species with which Donovan was acquainted, were in the collection of the late Duchess of Portland; one of which afterwards passed into the possession of Mr. Francillon.

LIBELLULA SERVILIA.

Plate 46. fig. 2.

CH. SP. L. alis hyalinis, basi flavis; corpore rubro. Expans. alar. $2\frac{3}{4}$ unc.
L. with the wings hyaline, yellow at the base, body red. Expanse of the wings $2\frac{3}{4}$ inches.
SYN. Libellula Servilia, *Drury Exot. Ent.* 1st edit. app. vol. 2.
Libellula ferrugata, *Fabr. Mant. Ins.* 336. 11.
Libellula ferruginea, *Fabr. Ent. Syst.* 2. *p.* 380. *Donovan,* 1st edit.

Very common in China.

LIBELLULA FULVIA.

Plate 46. fig. 3.

CH. SP. L. luteo-testacea, alis fulvescentibus, marginibus anticis testaceis maculâ mediâ subpellucidâ, stigmate ad apicem fusco. Expans. alar. $2\frac{1}{3}$ unc.
L. luteo-testaceous; wings fulvescent, with the anterior margins testaceous, having a pellucid spot in the middle, and the stigma brown. Expanse of the wings $2\frac{1}{3}$ inches.
SYN. Libellula Fulvia, *Drury Exot. Ent.* vol. 2. *pl.* 46. *fig.* 2.

1. Agrion Chinensis. 2. Libellula Serirlia.
3. Libellula Fulvia.

Epeira Maculata.

Order. DIMEROSOMATA. *Leach*.

EPEIRA (NEPHILA) MACULATA.
Plate 47.

Class.	Arachnida. *Lamarck.* Aptera p. *Linnæus.*
Order.	Dimerosomata, *Leach.*
Family.	Araneidæ, *Leach.*
Genus.	Epeira, *Walckenäer.*
Sub-Gen.	Nephila, *Leach Zool. Misc.*
Ch. Sp.	E. corpore elongato, cephalo-thorace holosericeo argenteo, abdomine cylindrico fusco-rubro lineis punctisque albis; pedibus longissimis atris. Long. Corp. 1¾ unc.
	E. with the body elongated, cephalo-thorax holosericeus and silvery, abdomen cylindric, red-brown with spots and lines of white, legs very long and black. Length of the body 1¾ inches.
Syn.	Aranea maculata, *Fabr. Ent. Syst.* 2. *p.* 425.

This remarkable creature is peculiar to some parts of the Chinese empire. It is not the largest of the genus known; yet it is of sufficient magnitude to excite terror and disgust. To an European, who has seen only the indigenous spiders of his own country, a species five or six inches in length, and nearly the same in breadth, must appear a frightful creature: Epeira Maculata sometimes exceeds that size; but it has not the forbidding aspect of most insects of the same genus. The legs are unusually long, and the body slender. In its general appearance it resembles some kinds of the *Phalangia* that are known in England by the vulgar name *Harvest-men*, being generally seen about that time of the year.

It has been observed, that nature oftentimes adorns the most deformed and loathsome of her creatures in the richest display of colours; and this is especially noticed in many sorts of snakes, toads, lizards, &c. Spiders seem also of this description: to a form the most hideous we frequently find united a brilliance of colours, and elegance of marking, that is scarcely excelled by any of the butterfly tribe,—the most beautiful of all lepidopterous insects. Our present subject is a striking proof of the latter part of this observation. The three figures in our plate of Epeira Maculata exhibit a front and a profile view of the insect, together with the front of the head at the third figure. The head is furnished with two very strong black mandibles, each terminated in an extremely acute point. The fore part of the cephalo-thorax, which is wholly of a fine silky appearance, and the colour of silver, bending over the mandibles in the form of an arch, or circular head-piece, gives it the resemblance of a black head with a crown of silver on the brow. This appearance is heightened in no small degree by three rugged prominences, one in the centre, and another on each side, on the upper part; and by the minute black eyes, which, like those of most spiders, sparkle with the lustre of small gems. These eyes are eight in number, four are placed immediately in the front of the silver-coloured circular front piece, and on each side are two placed close together thus : ∴ :

The body is really beautiful, the chief colour is deep brown, strongly tinged with bright purple; a broad stripe of orange colour passes down the abdomen from the cephalo-thorax to the apex: the whole is elegantly marked with a variety of cream-coloured lines and spots intersecting each other. Very little hair is found on any part of this spider except on the cephalo-thorax, which being rubbed off, discovers a hard testaceous black substance beneath.

The description given by Fabricius accords in every respect with our specimen. The only insect with which it could possibly be confounded is Aranea Pilipes, which also has never been figured; it differs, however, from Aranea Maculata in the very hairy clothing of the legs, and it has also two silver stripes down the back: a striking specific distinction to separate it from our insect. It is also a native of the East Indies, but not of China, that we are informed.

Order. DECAPODA. *Latreille*.

ORITHYIA MAMILLARIS.

Plate 48.

CLASS.	CRUSTACEA, *Cuvier*. Aptera p. *Linnæus*.
ORDER.	DECAPODA, *Latreille*.
SECTION.	BRACHYURA, *Latreille*.
FAMILY.	PORTUNIDÆ.
GENUS.	ORITHYIA. *Fabricius*. Cancer p. *Linn. Donov.*
CH. SP.	O. testâ utrinque trispinosâ, taberculatâ, maculis duabus rufis, fronte tridentato. Long. test. 1¼ unc.
	O. with the carapax having three spines on each side and tuberculated, with two red spots, the front tridentate. Length 1¼ inch.
SYN.	Cancer mamillaris, *Fabr. Ent. Syst.* 2. 465. 91.

It is worthy of remark, that this is the only species of the old genus *Cancer*, which Fabricius mentions as a native of China.

Orithyia mamillaris.

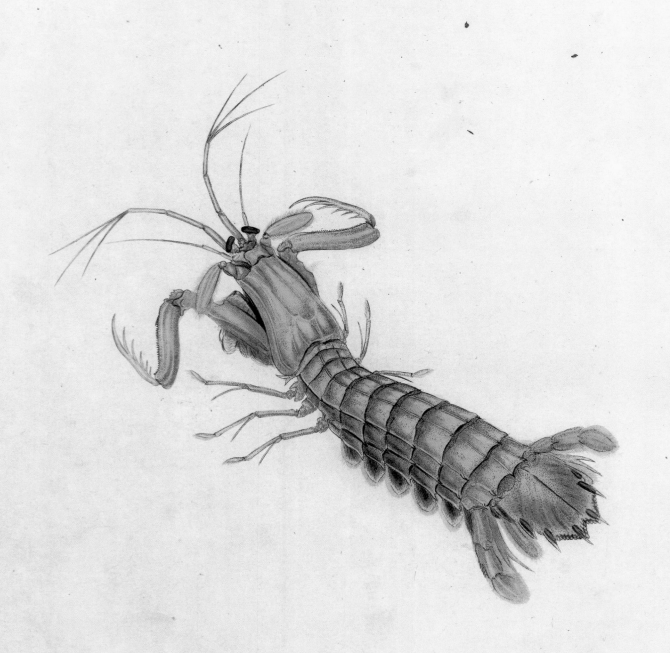

Squilla Mantis.

Order. STOMAPODA. *Latreille.*

SQUILLA MANTIS.

Plate 49.

Genus.	Squilla, *Fabricius.* Cancer, p. *Linnæus.*
Ch. Sp.	Sq. manibus compressis 6-dentatis, articulo ultimo abdominis carinâ centrali, dentibus tribus utrinque apiceque bidentato. Long. Corp. 4½ unc.
	Sq. with the terminal joint of the large pair of claws six toothed, the last abdominal segment with a central carina, three teeth on each side and two at the extremity. Length of the body 4½ inches.
Syn.	Cancer Mantis, *Linn. Syst. Nat.* 2. 1054. *Fab. Ent. Syst.* 2. 511. *Desmarest Cons. sur les Crustaces, p.* 251. *Encycl. Méth. pl.* 324. *De Geer Ins. vol.* 7. *t.* 34.

The Linnæan *Cancri* are numerous, and include many species not less singular in appearance than the extraordinary creature before us. Indeed, some species are so extremely different from the rest, both in structure and manners of life, that even Donovan could not hesitate in concluding the Linnæan character of the genus to be defective and indefinite. This may be observed in several of the species Linnæus himself described, and throughout a more extensive number of those discovered since the time of that author. It is evident Linnæus could never reconcile the subdivisions of the two principal families, the *Brachyuri* and *Macrouri*, or crabs with short and long tails; and later naturalists have ventured, with propriety, to alter this part of his arrangement.

Desmarest gives the Mediterranean as the locality of Squilla Mantis, but the species here figured, which is very common in the Chinese boxes of insects sent to this country, agrees with the extended characters given of S. Mantis. Fabricius says of it, " Habitat in mari *Asiatico, Indico, Mediterraneo*, Italis esculentus."

SCOLOPENDRA MORSITANS?

Plate 50.

CLASS.	AMETABOLA, *Leach.* Myriapoda, *Latreille.* Aptera p. *Linnæus.*
FAMILY.	SCOLOPENDRIDÆ.
GENUS.	SCOLOPENDRA, *Linnæus.*
CH. SP.	Sc. pedibus utrinque 21, posterioribus spinosis. Long. Corp. 6 unc. Sc. with twenty-one feet on each side, the posterior pair spined. Length of the body 6 inches.
SYN.	Scolopendra morsitans, *De Geer Ins. vol.* 7. *t.* 43. *fig.* 1. (in *Indiâ*). *Linn.? Syst. Nat.* 2. 1063. ("Hab. in *Indiis*"). *Fabricius Ent. Syst.* 2. *p.* 390. ("Hab. in *Indiâ orientali.*")

Travellers agree that the temperate parts of Asia would be a terrestrial paradise, were it not for the multitude of troublesome insects and reptiles with which they are infested. In a well cultivated country like China, many of these creatures can scarcely find shelter; but such as harbour in the walls or furniture of human dwellings are as abundant in that, as any other country lying within or near the tropics. Amongst the latter, none produce more terrible effects than the Centipede, whose poison is as venomous as that of the scorpion, which is also a native of China.

Sir G. Staunton mentions a remarkable circumstance that occurred during the embassy to China to which he was attached. The ambassador and his suite were accommodated in a temple near the suburbs of Tong-choo-foo. "In some of the apartments the priests had suffered scorpions and scolopendras to harbour through neglect. These noisome creatures were known only by description to some of the gentlemen in the embassy, who had not visited the southern parts of Europe: the sight of such, for the first time, excited a degree of horror in their minds; and it seemed to them to be a sufficient objection to the country, that it produced these animals." Sir George however adds, that no accident happened in that instance.—The species of Scolopendra he alludes to, is probably *Morsitans*, which is common in many parts of the world, but is particularly found of a frightful size, and in vast abundance, in the two Indies.

Many authors have described this creature. In the voluminous works of Seba we find several specimens of it from different countries, differing materially in size, and some trifling particulars. The largest of these exceed our figure in magnitude, being near fourteen inches in length: this he calls *Millepeda major* ex nova Hispania. His figure of *Millepeda Africana* is about the size of our Chinese specimen. He has also a third

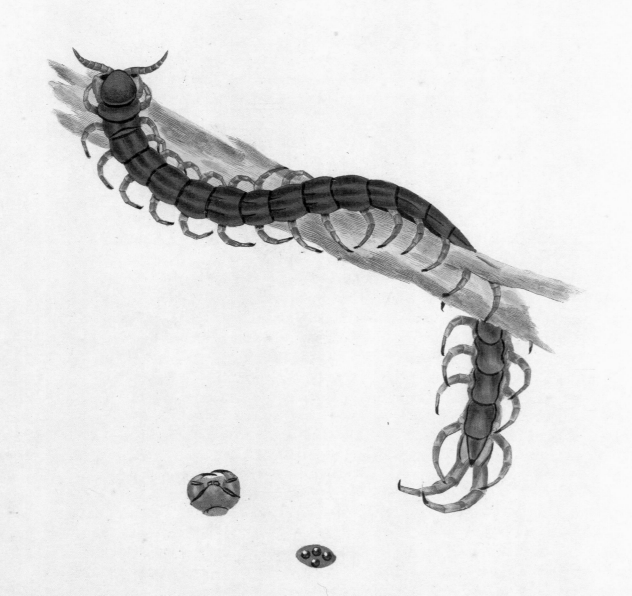

Scolopendra Morsitans.

MYRIAPODA.

and fourth figure, *Millepeda Orientalis* and *Millepeda Ceylonica, mas:* the latter is the same length as our figure, but the body is very narrow. Millepeda Orientalis is also the same length, but the body is very broad. Some of these insects are not four inches in length. These will be regarded as distinct species; and, indeed, it is questionable whether the Chinese species be strictly entitled to the specific name of Morsitans, as Guérin (Encyclop. Méth. x. p. 393), and Pohl and Kollar, in their work on the noxious insects of Brazil, have given the name of Morsitans to the Brazilian species, which has also twenty-one pairs of legs; to this species, however, Dr. Leach gave the specific name of Alternans. The entire genus has, indeed, need of a monographical revision. It will be seen that the antennæ in this figure are much shorter than in that of De Geer, &c.

Authors agree that they vary exceedingly in size* and colour. De Geer describes them to be sometimes deep reddish brown; at others, the colour of yellow ochre. The figure in Catesby's Natural History of Carolina is light brown; we have specimens of a livid yellow, and have seen others strongly tinged with red.

The last pair of legs is considerably larger than the others, and is armed with small black spines. The legs terminate in very sharp hooks or nails of a shining black colour. All the other legs are also furnished with a smaller nail of the same shape and colour.

M. Gronovius says, that all its feet are very venomous; but the most formidable of its weapons are the two sharp hooked instruments that are placed under the mouth, and with which it destroys its prey.

Leuwenhoeck having examined these instruments with a microscope, found a small opening at the extremity of each, and a channel from them into the body of the creature. Through this channel he supposes the Scolopendra emits the poisonous fluid into the wound it makes with the hooked instrument. That author further remarks, that he has seen a liquor on that part of living scolopendras. A figure of these instruments on the under side of the head is represented in one of the dissections in our plate.

The same author, wishing to ascertain the influence of the poison of Scolopendra morsitans, placed a large fly within its reach. The Scolopendra at first took it between a pair of its middle feet, then passed it from one pair of feet to the next, till the fly was brought under the sharp pointed instrument or crotchets at the head, which it plunged into the fly, and it died instantly. Notwithstanding this experiment, De Geer, Catesby, and other authors assert, that its bite seldom proves fatal to larger animals; but all agree

---

* These creatures differ from most insects in their manner of growth, insomuch that it is impossible to ascertain when they are of their full size. The segments of the body increase in number as they advance in age, which circumstance renders it difficult oftentimes to determine the species without a minute examination of its other characters.

N

90   MYRIAPODA.

that its poison is as dangerous as that of the scorpion. (See also Worbe in the Bulletin de la Soc. Philomat, Jan. 1824, Amoreux Insectes venimeux, and the recent work of Pohl and Kollar above referred to, for further details relative to the poisonous properties of these insects.)

This Scolopendra has eight eyes: they are very small; four are placed on each side of the head near the antennæ. In the dissections a figure is given to exhibit the manner in which the four eyes are placed on one side.

ALPHABETICAL INDEX.

In this Index the names employed both in the present and former editions are introduced, in order to render the references which have been made by writers to the former edition available. The names, generic, sub-generic, or specific, first employed in the present edition are distinguished by a *.

*Acræa Vesta, Pl. 30. f. 1.
*Agrion Chinensis, Pl. 46. f. 1.
Aranea maculata, Pl. 47.
*Argynnis Erymanthis, Pl. 35. f. 1.
*Astemma Schlanbuschii, Pl. 20. f. 2.
*Æshna clavata, Pl. 45. f. 1.
*Belostoma *Indica? Pl. 18.
*Belostoma *(Sphærodema) rustica, Pl. 19. f. 1.
Buprestis ocellata, Pl. 7. f. 2.
Buprestis *(Chrysochroa) ocellata, Pl. 7. f. 2.
Buprestis vittata, Pl. 7. f. 1.
Buprestis *(Chrysochroa) vittata, Pl. 7. f. 1.
*Calandra longipes, Pl. 4. f. 2.
*Callidea *ocellata, Pl. 20. f. 1.
*Callidea Stockerus, Pl. 21. f. 1.
*Callimorpha? bifasciata, Pl. 41. f. 4.
*Callimorpha? *panthorea, Pl. 44. f. 3.
*Callimorpha? ruficollis, Pl. 41. f. 3.
*Callimorpha? Thallo, Pl. 41. f. 2.
Cancer Mamillaris, Pl. 48.
Cancer Mantis, Pl. 49.
Cerambyx farinosus, Pl. 6. f. 3.
Cerambyx reticulator, Pl. 6. f. 2.
Cerambyx Rubus, Pl. 6. f. 1.
*Cerbus *tenebrosus? Pl. 21. f. 4.
*Cercopis abdominalis, Pl. 16. f. 5.
Cetonia Chinensis, Pl. 3. f. 1.
Cetonia *(Tetragona) Chinensis, Pl. 3. f. 1.
*Chariesterus cruciger, Pl. 21. f. 3.
Cicada abdominalis, Pl. 16. f. 5.
Cicada ambigua, Pl. 16. f. 2.
Cicada atrata, Pl. 15.
Cicada frontalis, Pl. 16. f. 6.

Cicada lanata, Pl. 16. f. 3.
Cicada limbata var. Pl. 17.
Cicada sanguinea, Pl. 16. f. 1.
Cimex aurantius, Pl. 21. f. 2.
Cimex bifidus, Pl. 21. f. 5.
Cimex cruciger, Pl. 21. f. 3.
Cimex dispar, Pl. 20. f. 1.
Cimex Phasianus, Pl. 21. f. 4.
Cimex Slanbuschii, Pl. 20. f. 2.
Cimex Stockerus, Pl. 21. f. 1.
*Cleonis perlatus, Pl. 4. f. 7.
*Colias *(Callidryas) Philea, Pl. 32. f. 2.
*Colias *(Callidryas) Pyranthe, Pl. 32. f. 1.
*Copris Bucephalus, Pl. 2. f. 3.
*Copris Midas, Pl. 1. f. 1.
*Copris Molossus, Pl. 2. f. 1.
Curculio barbirostris, Pl. 4. f. 3.
Curculio Chinensis, Pl. 4. f. 1.
Curculio longipes, Pl. 4. f. 2.
Curculio perlatus, Pl. 4. f. 7.
Curculio pulverulentus, Pl. 4. f. 6.
Curculio squamosus, Pl. 4. f. 5. and Pl. 5.
Curculio verrucosus, Pl. 4. f. 4.
*Cynthia Almana, Pl. 36. f. 2.
*Cynthia Œnone, Pl. 36. f. 1.
*Cynthia Orithya, Pl. 35. f. 2.
*Deilephila Nechus, Pl. 40. f. 1.
*Epeira *(Nephila) maculata, Pl. 47.
*Erebus *macrops, Pl. 44. f. 1.
*Euchlora viridis, Pl. 3. f. 2.
*Eusemia lectrix, Pl. 43. f. 2.
*Flata *nigricornis, Pl. 17.
Fulgora Candelaria, Pl. 14.

*Glaucopis Polymena, Pl. 40. f. 2.
*Gryllotalpa *Chinensis, Pl. 12. f. 2.
Gryllus acuminatus, Pl. 11. f. 2.
Gryllus *(Conocephalus) acuminatus? Pl. 11. f. 2.
Gryllus flavicornis, Pl. 12. f. 1.
Gryllus Gryllotalpa, Pl. 12. f. 2.
Gryllus morbillosus, Pl. 13.
Gryllus nasutus, Pl. 10. f. 1.
Gryllus perspicillatus, Pl. 11. f. 1.
Gryllus *(Phasgoneurus) perspicillatus, Pl. 11. f. 1.
Gryllus vittatus, Pl. 10. f. 2.
*Gymnopleurus *sinuatus, Pl. 1. f. 5.
*Harpactor bifidus, Pl. 21. f. 5.
*Heleona militaris, Pl. 43. f. 1.
*Hipparchus Zonarius, Pl. 44. f. 2.
*Hipporhinus verrucosus, Pl. 4. f. 4.
*Hypomeces squamosus, Pl. 4. f. 5. and Pl. 5.
*Hypomeces squamosus *var. Pl. 4. f. 6.
*Lamia *punctator, Pl. 6. f. 3.
*Lamia reticulator, Pl. 6. f. 2.
*Lamia Rubus, Pl. 6. f. 1.
Libellula Chinensis, Pl. 46. f. 1.
Libellula clavata, Pl. 45. f. 1.
Libellula ferruginea, Pl. 46. f. 2.
Libellula Fulvia, Pl. 46. f. 3.
Libellula Indica, Pl. 45. f. 2.
Libellula *Servilia, Pl. 46. f. 2.
Libellula 6-maculata, Pl. 45. f. 3.
Libellula *variegata, Pl. 45. f. 2.
*Limenitis *Eurynome, Pl. 35. f. 4.
*Limenitis *Leucothoe, Pl. 35. f. 3.
*Locusta *(Rutidoderes) flavicornis, Pl. 12. f. 1.
*Locusta *(Phymatea) morbillosa, Pl. 13.
*Lystra lanata, Pl. 16. f. 3.
Mantis *(Schizocephala) *bicornis, Pl. 9. f. 1.
Mantis flabellicornis, Pl. 9. f. 2.
Mantis *(Empusa) flabellicornis, Pl. 9. f. 2.
Mantis oculata, Pl. 9. f. 1.
Meloe Cichorei, Pl. 8. f. 1.
Melolontha viridis, Pl. 3. f. 2.
*Morpho *(Drusilla) Jairus, Pl. 33.
*Morpho Rhetenor, Pl. 29.
*Mylabris Cichorii, Pl. 8. f. 1.
*Myrina *(Loxura) Atymnus, Pl. 39. f. 1.
Nepa grandis, Pl. 18.
Nepa rubra, Pl. 19. f. 2.
Nepa rustica, Pl. 19. f. 1.

*Nymphalis Antiochus, Pl. 37. f. 2.
*Nymphalis *(Charaxes) Bernardus, Pl. 34.
*Nymphalis Jacintha, Pl. 37. f. 1.
*Nymphalis *(Aconthea) Lubentina, Pl. 36. f. 3.
*Nymphalis *Sylla, Pl. 38.
*Oniticellus cinctus, Pl. 1. f. 3.
*Onthophagus seniculus, Pl. 2. f. 2.
*Orithyia mamillaris, Pl. 48.
*Oryctes *Rhinoceros? Pl. 1. f. 2.
Papilio Agamemnon, Pl. 26. f. 2.
Papilio Agenor, Pl. 24. f. 2.
Papilio Almana, Pl. 36. f. 2.
Papilio Antiochus, Pl. 37. f. 2.
Papilio Atymnus, Pl. 39. f. 1.
Papilio Bernardus, Pl. 34.
Papilio Coon, Pl. 24. f. 1.
Papilio Crino, Pl. 23.
Papilio Demoleus, Pl. 28. f. 2.
Papilio Epius, Pl. 28. f. 1.
Papilio Erymanthis, Pl. 35. f. 1.
Papilio Gambrisius, Pl. 38.
Papilio Glaucippe, Pl. 31. f. 1.
Papilio Hyparete, Pl. 30. f. 3.
Papilio Jacintha, Pl. 37. f. 1.
Papilio Jairus, Pl. 33.
Papilio Laomedon, Pl. 27.
Papilio Leucothoe, Pl. 35. f. 4.
Papilio Lubentina, Pl. 36. f. 3.
Papilio Mæcenas, Pl. 39. f. 2.
Papilio Œnone, Pl. 36. f. 1.
Papilio Orythia, Pl. 35. f. 2.
Papilio Paris, Pl. 22.
Papilio Pasithoe, Pl. 30. f. 2.
Papilio Peranthus, Pl. 25.
Papilio Philea, Pl. 32. f. 2.
Papilio Polyxena, Pl. 35. f. 3.
Papilio *Protenor, Pl. 27.
Papilio Pyranthe, Pl. 32. f. 1.
Papilio Rhetenor, Pl. 29.
Papilio Sesia, Pl. 31. f. 2.
Papilio Telamon, Pl. 26. f. 1.
Papilio Vesta, Pl. 30. f. 1.
Phalæna Atlas, Pl. 42.
Phalæna bubo, Pl. 44. f. 1.
Phalæna lectrix, Pl. 43. f. 2.
Phalæna militaris, Pl. 43. f. 1.
Phalæna pagaria, Pl. 44. f. 3.

ALPHABETICAL INDEX.

Phalæna zonaria, Pl. 44. f. 2.
*Pieris *(Iphias) Glaucippe, Pl. 31. f. 1.
*Pieris Hyparete, Pl. 30. f. 3.
*Pieris Pasithoe, Pl. 30. f. 2.
*Pieris *(Thestias) *Pyrene? Pl. 31. f. 2.
*Raphigaster aurantius, Pl. 21. f. 2.
*Rhina barbirostris, Pl. 4. f. 3.
*Rhinastus *sternicornis, Pl. 4. f. 1.
*Sagra *splendida, Pl. 8. f. 2.
*Saturnia Atlas, Pl. 42.
Scarabæus Bucephalus, Pl. 2. f. 3.
Scarabæus cinctus, Pl. 1. f. 3.
Scarabæus Leei, Pl. 1. f. 5.
Scarabæus Midas, Pl. 1. f. 1.
Scarabæus Molossus, Pl. 2. f. 1.
Scarabæus nasicornis, Pl. 1. f. 2.
Scarabæus sacer, Pl. 1. f. 4.
Scarabæus *(Heliocantharus) *sanctus? Pl. 1. f. 4.
Scarabæus seniculus, Pl. 2. f. 2.
Scolopendra morsitans, Pl. 50.
*Sesia Hylas, Pl. 41. f. 1.
Sphinx bifasciata, Pl. 41. f. 4.
Sphinx Hylas, Pl. 41. f. 1.
Sphinx Nechus, Pl. 40. f. 1.
Sphinx Polymena, Pl. 40. f. 2.
Sphinx ruficollis, Pl. 41. f. 3.
Sphinx Thallo, Pl. 41. f. 2.
*Squilla mantis, Pl. 49.
Tenebrio femoratus, Pl. 8. f. 2.
*Tettigonia frontalis, Pl. 16. f. 6.
Tettigonia splendidula, Pl. 16. f. 4.
*Thecla Mæcenas, Pl. 39. f. 2.
*Truxalis *Chinensis, Pl. 10. f. 1.
*Truxalis *(Mesops) vittatus, Pl. 10. f. 2.

SYSTEMATIC INDEX.

INSECTA.

I.—MOUTH WITH JAWS.

Order. COLEOPTERA.

Tribe. LAMELLICORNES, *Latr.*
Family. SCARABÆIDÆ, *Mac L.*

Scarabæus (Heliocantharus) sanctus? Pl. 1. f. 4.
Gymnopleurus sinuatus, Pl. 1. f. 5.
Copris Midas, Pl. 1. f. 1.
C.   Molossus, Pl. 2. f. 1.
C.   Bucephalus, Pl. 2. f. 3.
Onthophagus seniculus, Pl. 2. f. 2.
Oniticellus cinctus, Pl. 1. f. 3.

Family. DYNASTIDÆ, *Mac L.*

Oryctes Rhinoceros, Pl. 1. f. 2.

Family. MELOLONTHIDÆ, *Mac L.*

Euchlora viridis, Pl. 3. f. 2.

Family. CETONIIDÆ, *Mac L.*

Cetonia (Tetragona) Chinensis, Pl. 3. f. 1.

Tribe. STERNOXI, *Latr.*
Family. BUPRESTIDÆ, *Leach.*

Buprestis (Chrysochroa) vittata, Pl. 7. f. 1.
——————————— ocellata, Pl. 7. f. 2.

Tribe. TRACHELIDES, *Latr.*
Family. MELOIDÆ, *Westw.*

Mylabris Cichorii, Pl. 8. f. 1.

Tribe. RHYNCOPHORA, *Latr.*
Family. CURCULIONIDÆ, *Leach.*

Hipporhinus verrucosus, Pl. 4. f. 4.
Hypomeces squamosus, Pl. 4. f. 5. and Pl. 4. f. 6.
——————————— var. Pl. 5.

Cleonis perlatus, Pl. 4. f. 7.
Rhinastus sternicornis, Pl. 4. f. 1.
Rhina barbirostris, Pl. 4. f. 3.
Calandra longipes, Pl. 4. f. 2.

Tribe. LONGICORNES, *Latr.*
Family. LAMIIDÆ, *Westw.*

Lamia Rubus, Pl. 6. f. 1.
L.   reticulator, Pl. 6. f. 2.
L.   punctator, Pl. 6. f. 3.

Tribe. EMPODA.
Family. SAGRIDÆ, *Westw.*

Sagra splendida, Pl. 8. f. 2.

———

Order. ORTHOPTERA. *Oliv.*

Tribe. CURSORIA, *Latr.*
Family. MANTIDÆ, *Leach.*

Mantis (Empusa) flabellicornis, Pl. 9. f. 2.
M. (Schizocephala) bicornis, Pl. 9. f. 1.

Tribe. SALTATORIA, *Latr.*
Family. LOCUSTIDÆ, *Leach.*

Truxalis Chinensis, Pl. 10. f. 1.
T.   (Mesops) vittatus, Pl. 10. f. 2.
Locusta (Rutidoderes) flavicornis, Pl. 12. f. 1.
L.   (Phymatea) morbillosa, Pl. 13.

Family. GRYLLIDÆ, *Leach.*

Gryllus (Phasgonurus) perspicillatus, Pl. 11. f. 1.
G.   (Conocephalus) acuminatus? Pl. 11. f. 2.

Family. ACHETIDÆ, *Leach.*

Gryllotalpa Chinensis, Pl. 12. f. 2.

SYSTEMATIC INDEX.

Order. NEUROPTERA. *Linn.*
Family. LIBELLULIDÆ, *Leach.*

Æshna clavata, Pl. 45. f. 1.
Libellula variegata, Pl. 45. f. 2.
Libellula 6-maculata, Pl. 45. f. 3.
L.   Servilia, Pl. 46. f. 2.
L.   Fulvia, Pl. 46. f. 3.
Agrion Chinensis, Pl. 46. f. 1.

II.—MOUTH SUCTORIAL.

Order. LEPIDOPTERA. *Linn.*
Tribe. DIURNA, *Latr.*
Family. PAPILIONIDÆ, *Leach.*

Papilio Paris, Pl. 22.
P.   Crino, Pl. 23.
P.   Coon, Pl. 24. f. 1.
P.   Agenor, Pl. 24. f. 2.
P.   Peranthus, Pl. 25.
P.   Telamon, Pl. 26. f. 1.
P.   Agamemnon, Pl. 26. f. 2.
P.   Protenor ♀, Pl. 27.
P.   Epius, Pl. 28. f. 1.
P.   Demoleus, Pl. 28. f. 2.
Pieris Pasithoe, Pl. 30. f. 2.
P.   Hyparete, Pl. 30. f. 3.
P.   (Iphias) Glaucippe, Pl. 31. f. 1.
P.   (Thestias) Pyrene? Pl. 31. f. 2.
Colias (Callidryas) Pyranthe, Pl. 32. f. 1.
C.   (C.   ) Philea, Pl. 32. f. 2.

Family. HELICONIIDÆ, *Swainson.*

Acræa Vesta, Pl. 30. f. 1.

Family. NYMPHALIDÆ, *Swainson.*

Morpho Rhetenor, Pl. 29.
M.   (Drusilla) Jairus, Pl. 33.
Nymphalis Jacintha, Pl. 37. f. 1,
N.   Antiochus, Pl. 37. f. 2.
N.   Sylla, Pl. 38.
N.   (Charaxes) Bernardus, Pl. 34.
N.   (Aconthea) Lubentina, Pl. 36. f. 3.
Argynnis Erymanthis, Pl. 35. f. 1.
Cynthia Orithya, Pl. 35. f. 2.
C.   Œnone, Pl. 36. f. 1.
C.   Almana, Pl. 36. f. 2.
Limenitis Leucothoe, Pl. 35. f. 3.
L.   Eurynome, Pl. 35. f. 4.

Family. LYCÆNIDÆ, *Swainson.*

Myrina (Loxura) Atymnus, Pl. 39. f. 1.
Thecla Mæcenas, Pl. 39. f. 2.

Section. CREPUSCULARIA, *Latr.*
Family. SPHINGIDÆ, *Leach.*

Deilephila Nechus? Pl. 40. f. 1.
Sesia Hylas, Pl. 41. f. 1.

Section. ————?
Family. ZYÆNIDÆ, *Leach.*

Glaucopis Polymena, Pl. 40. f. 2.

Section. NOCTURNA, *Latr.*
Family. BOMBYCIDÆ, *Leach.*

Saturnia Atlas, Pl. 42.
Eusemia Lectrix, Pl. 43. f. 2.

Family. ARCTIIDÆ, *Stephens.*

Callimorpha? Thallo, Pl. 41. f. 2.
C.   ? ruficollis, Pl. 41. f. 3.
C.   ? bifasciata, Pl. 41. f. 4.
C.   ? panthorea, Pl. 44. f. 3.
Heleona militaris, Pl. 43. f. 1.

Family. NOCTUIDÆ, *Leach.*

Erebus macrops, Pl. 44. f. 1.

Family. GEOMETRIDÆ, *Leach.*

Hipparchus Zonarius, Pl. 44. f. 2.

Order. HEMIPTERA. *Linn.*
Sub-Order. HETEROPTERA, *Latr.*
Section. GEOCORISA, *Latr.*
Family. SCUTELLERIDÆ, *Leach.*

Callidea ocellata, Pl. 20. f. 1.
C.   Stockerus, Pl. 21. f. 1.
Raphigaster aurantius, Pl. 21. f. 2.

SYSTEMATIC INDEX.

Family. COREIDÆ, *Leach.*
Chariesterus cruciger, Pl. 21. f. 3.
Cerbus tenebrosus? Pl. 21. f. 4.

Family. LYGÆIDÆ, *Leach.*
Astemma Schlanbuschii, Pl. 20. f. 2.

Family. REDUVIIDÆ, *Leach.*
Harpactor bifidus, Pl. 21. f. 5.

Section. HYDROCORISA, *Latr.*
Family. NEPIDÆ, *Leach.*
Belostoma Indica? Pl. 18.
B. (Sphærodema) rustica, Pl. 19. f. 1.
Nepa rubra, Pl. 19. f. 2.

Sub-Order. HOMOPTERA, *Latr.*
Family. CICADIDÆ, *Leach.*
Cicada atrata, Pl. 15.
C. sanguinea, Pl. 16. f. 1.
C. ambigua, Pl. 16. f. 2.
C. splendidula, Pl. 16. f. 4.

Family. FULGORIDÆ, *Leach.*
Fulgora Candelaria, Pl. 14.
Lystra lanata, Pl. 16. f. 3.
Flata nigricornis, Pl. 17.

Family. CERCOPIDÆ, *Leach.*
Cercopis abdominalis, Pl. 16. f. 5.
Tettigonia frontalis, Pl. 16. f. 6.

CRUSTACEA.

Order. DECAPODA.
Orithyia mamillaris, Pl. 48.

Order. STOMAPODA.
Squilla mantis, Pl. 49.

ARACHNIDA.

Epeira (Nephila) clavipes, Pl. 47.

AMETABOLA.

Scolopendra morsitans, Pl. 50.

THE END.

C. Whittingham, Tooks Court, Chancery Lane.

图书在版编目（CIP）数据

中国昆虫志/（英）爱德华·多诺万著；
（英）约翰·韦斯特伍德修订；徐锦华主编.
-- 上海：上海古籍出版社，2023.6
（徐家汇藏书楼珍稀文献选刊）
ISBN 978-7-5732-0616-9
Ⅰ.①中… Ⅱ.①爱…②约…③徐… Ⅲ.①昆虫志
—中国 Ⅳ.①Q968.22
中国国家版本馆CIP数据核字（2023）第032058号

丛书主编：徐锦华
丛书总序：董少新
本册导言：邓　岚

责任编辑：虞桑玲
装帧设计：严克勤
技术编辑：隗婷婷

徐家汇藏书楼珍稀文献选刊
中国昆虫志
Natural History of the Insects of China

［英］爱德华·多诺万（Edward Donovan）著
［英］约翰·韦斯特伍德（John Obadiah Westwood）修订

上海古籍出版社出版发行
（上海市闵行区号景路159弄1-5号A座5F　邮政编码201101）
　（1）网址：www.guji.com.cn
　（2）E-mail: guji@guji.com.cn
　（3）易文网网址：www.ewen.co

印刷：上海丽佳制版印刷有限公司

开本：787×1092毫米　1/8
插页：5　印张：24.5　　　字数：190千字
版次：2023年6月第1版　　2023年6月第1次印刷
ISBN　978-7-5732-0616-9/J·676
定价：498.00元